COMPUTER PROGRAMMING IN BASIC

Holden-Day Computer and Information Sciences Series

S.D. Conte, Editor

HOLDEN-DAY

San Francisco, Cambridge, London, Amsterdam

Harold Sunderland

COMPUTER PROGRAMMING IN BASIC

JOSEPH P. PAVLOVICH
THOMAS E. TAHAN

First Printing, June, 1971
Second Printing, August, 1971

Copyright © 1971 by Holden-Day, Inc., San Francisco, Calif. All rights reserved. No part of this publication may be reproduced in any form, stored in a retrieval system, or transmitted by any means without prior written permission of the publisher.

Printed in the United States of America

Library of Congress Catalog Card Number: 73-155558

ISBN: 0-8162-6653-0

FOREWORD

Holden-Day, Inc. is pleased to publish this textbook in its Computer and Information Sciences Series. It is the most thorough treatment of the BASIC language available in textbook form today. The popularity of the BASIC language rests primarily on its simplicity and ease of use. In spite of this simplicity the authors show that it can also be used to treat very sophisticated and complex problems. In this book BASIC emerges as a language as powerful as FORTRAN, yet much simpler to learn and use. It should serve as an ideal textbook for an introductory course in computer programming for students who have had a traditional high school background in mathematics.

                        S.D. Conte
                        Editor
                        Computer and Information Sciences Series

PREFACE

This book introduces and develops a full repertoire of problem-solving techniques that can be used on any Time-Sharing System using the BASIC language. It is also suitable for use on a batch system which utilizes BASIC. The book leads the reader systematically through a set of problems which introduce and develop all the features of the BASIC language. By using this book in conjunction with a computer, you can learn how to program completely on your own, and at your own pace. The book can also be used as the text for a one-semester course on computer programming at the first or second year college level or even at the advanced high school level.

This material was developed and tested on four different Time-Sharing Systems: the PDP-10 computer at the Digital Equipment Corporation, Maynard, Massachusetts; the Harvard University SDS 940 Time-Sharing System, Cambridge; the Call-A-Computer Time-Sharing System, Boston; and the Multiplexed Information and Computing Service (Multics) at the Massachusetts Institute of Technology, Cambridge. Although the programs are written in the BASIC language, most features of the problem-solving algorithms can be implemented in other languages.

The problems presented in this book are mainly mathematical in nature, and are directed toward students with varying degrees of mathematical background: from one year of algebra through one year of calculus. These problems are used to demonstrate useful programming techniques. The range of other problems to which these problem-solving techniques may be applied is limited only by one's imagination.

This book is written from a problem-solving point of view. To program the solution to any problem, it is first necessary to understand thoroughly the characteristics of the problem. Thus it is to be expected that a significant by-product of learning how to program the solution to mathematical problems is a greater appreciation for and understanding of the mathematical ideas which underlie the problem itself.

The authors are indebted to many people who have made a contribution to this book: the Digital Equipment Corporation and Mr. Kenneth Olsen, its President, for giving us free time on their PDP-10 computer; Mr. Richmond Mayo-Smith, Headmaster of the Roxbury Latin School, who gave his support to this project; Mr. Richard Gumpertz, who made some valuable suggestions; and Mrs. Essad Tahan, for typing the manuscript. Also, although Thomas Tahan is still a student and a bachelor, Joseph Pavlovich is not, and is indebted to his wife Lurline for her patience and understanding.

J.P.P.
T.E.T.

4/71

CONTENTS

CHAPTER 1   1

1-1 What Is Time-Sharing? 1;   1-2 Purpose 2.

CHAPTER 2   3

2-1 The Teletype 3;   2-2 Punching a Tape 3;   2-3 LOGging ON 6; 2-4 Reading In a Punched Tape 7;   2-5 LOGging OFF 8.

CHAPTER 3   10

3-1 Command PRINT 10;
    1. Line Numbers 10;   2. Spaces 11;   3. Execution of a Program 11;   4. END Statement 12;   5. Replacing a Line 12;   6. Quotation Marks 13;   7. Commas with Numbers 14;   8. Commas with Strings 14;   9. Commas with Numbers and Strings 15;   10. Skipping a Line 16;   11. Semicolon with Numbers 17;   12. Semicolon with Strings 17;   13. PRINT TAB(N) Command 18;   14. Legal Expressions in PRINT Statements 19;   15. Exponential Notation 19;   16. ERROR Messages 21;

3-2 Operations $\uparrow, *, /, +, -$ 22;
    1. Symbols Used for Arithmetic Operations 22;   2. Order of Operations 22;   3. Use of Parentheses 23;

3-3 Corrections 25;
    1. SHIFT O Key (Back Arrow) 25;   2. Retyping a Line 26;   3. ALTMODE (ESCAPE or PREFIX) Key 27;

3-4 Command LET 29;
    1. Assignment of Values to Variables 29;   2. Legal Expressions in LET Statements 31;   3. Legal Names for Variables 31;   4. The Same Variable on Both Sides of an Equal Sign 33;   5. The Assignment of the Same Value to Several Variables 33;   6. The Word LET Omitted 34;

3-5 READ and DATA Statements 35;
  1. Method of Assigning Values 35; 2. The Order of Assignment of Values 35; 3. The Placement of READ Statements 37; 4. Legal Numbers in DATA Statements 37; 5. Insufficient DATA 38; 6. The Command RESTORE 38;
3-6 The INPUT Command 41;
  1. Using the INPUT Command 41; 2. INPUTting Several Variables 41; 3. Legal Numbers in INPUT Statements 43;
3-7 Summary 44.

CHAPTER 4   45

4-1 Built-in Functions 45;
  1. What Is a Function? 45; 2. Functions in BASIC 46; 3. Absolute Value and Sign Functions 47; 4. The Greatest Integer Function; Rounding Off 47; 5. The Random Number Function 49; 6. The Square Root Function 51; 7. The Logarithmic Function 52; 8. The Exponential Function 54; 9. Trigonometric Functions 54;
4-2 Numbers in BASIC 57;
  1. Numbers Accepted in BASIC 57; 2. Range and Accuracy 57; 3. OUTPUT Numbers 58;
4-3 Summary 59.

CHAPTER 5   60

5-1 Making a Flow Chart 60;
5-2 Program (1): The Area of a Circle and the Volume of a Sphere (1) 62;
  1. Program (1) 62; 2. Comments about Program (1) 63; 3. The Execution of a Program 64; 4. The Procedure after the Execution of a Program 64;
5-3 The IF-THEN Command; Program (2): The Area of a Circle and the Volume of a Sphere (2) 67;
  1. The IF-THEN Command 67; 2. The Compilation of a Program 68; 3. The Flow Chart of Program (2) 68; 4. Program (2): The Area of a Circle and the Volume of a Sphere (2) 69; 5. Comments about Program (2) 70;
5-4 Alphanumeric Data and String Variables 71;
  1. Strings and String Variables 71; 2. Renaming a String Variable 72; 3. READ-DATA and INPUT Statements with String Variables

Contents ix

73; 4. Mixing Numeric and String Data 74; 5. Command RESTORE with Alphanumeric Data 75; 6. The IF-THEN Command with String Variables 75; 7. The Comparison of String Variables 76; 8. Program (3): The Area of a Circle and the Volume of a Sphere (3) 78;

5-5 The GO TO Command; Program (4): The Real Roots of $AX^2 + BX + C = 0$ 79;
1. The GO TO Command 79; 2. The Real Roots of $AX^2 + BX + C = 0$ 79; 3. The Flow Chart of Program (4) 79; 4. Program (4): The Real Roots of $AX^2 + BX + C = 0$ 81; 5. Comments about Program (4) 81; 6. The ON-GO TO Command 83;

5-6 Loops; Program (5): N! (Factorial) 85;
1. The Loop 85; 2. N! (Factorial) 87; 3. The Flow Chart for N! (Factorial) 87; 4. Program (5): N! (Factorial) 89; 5. Comments about Program (5) 90;

5-7 The FOR-NEXT Command; Program (6): N!, A Second Approach 91;
1. Constructing a Loop with the FOR-NEXT Command 91; 2. Nested Loops 96; 3. The Transfer of Control outside the Loop 97; 4. Flow Chart for Program (6): N!, A Second Approach 98; 5. Program (6): N!, A Second Approach 98; 6. Comments about Program (6) 100;

5-8 The Command DIM 101;
1. Its Use with Subscripted Variables 101; 2. More than One Variable 103; 3. A DIM Statement before a Variable 103; 4. Subscripted String Variables 104;

5-9 Summary 106.

CHAPTER 6 107

6-1 The GOSUB-RETURN Commands; Program (7): A Dice Game 107;
1. Subroutine 107; 2. Commands GOSUB-RETURN 107; 3. The Distinction between GOSUB and GO TO Commands 110; 4. Comments about Program (7) 111; 5. Facts about the GOSUB-RETURN Commands 112;

6-2 The MAT INPUT, NUM Commands; Program (8): Sum, Product, Maximum, Minimum 113;
1. MAT INPUT and NUM 113; 2. Comments about Program (8) 116;

6-3 The CHANGE Command; Program (9): Algebra Quiz 118;
1. The ASCII Code 118; 2. The CHANGE Command 118; 3. ASCII Code Numbers 120; 4. Other Useful ASCII Code Numbers 122; 5. Extracting Characters from a String 122; 6. Program (9): Algebra Quiz 124; 7. Comments about Program (9) 126;

6-4 Defining Functions 130;
   1. The Value of a Function 130;   2. The Command DEF 131;
   3. Characteristics of User-defined Functions 133;   4. Multiline
   Function Definitions 134;   5. Functions and Subroutines 138;
6-5 Program (10). To Find the Slope of F(X) at X = C 139;
   1. The Slope of F(X) at X = C 140;   2. Using the Computer to Find
   the Slope Number 142;   3. Program (10) 143;   4. Comments
   about Program (10) 143;
6-6 Summary 146.

CHAPTER 7   148

7-1 Techniques of Debugging 148;
   1. Bugs 148;   2. ERROR Messages 148;   3. Errors in the Logical
   Progression of Statements 152;   4. General Techniques of Debugging 154;
7-2 Program (11): Solutions of Triangles in Trigonometry 154;
   1. Case I: Angle-Side-Angle 154;   2. Case II: Side-Angle-Side 155;
   3. Case III: Side-Side-Side 157;   4. Case IV: Side-Side-Angle
   (Ambiguous Case) 159;   5. Traditional Solution of Triangle
   Problem 164;   6. Program (11) 164;   7. Comments about
   Program (11) 169;
7-3 Program (12): The Graph of Any Function 171;
   1. Rectangular Coordinate System 171;   2. Graph by Computer
   173;   3. Program (12) 174;   4. Comments about Program (12)
   176;
7-4 Program (13): Real Zeros of Any Function 180;
   1. Zeros of a Function 180;   2. Continuous Function 180;
   3. Finding Zeros by the Computer 181;   4. Program (13) 181;
   5. Comments about Program (13) 183;
7-5 Summary 187.

CHAPTER 8   188

8-1 Matrices 188;
   1. What Is a Matrix? 188;   2. The Sum and Scalar Product of
   Matrices 189;   3. The Zero Matrix, Additive Inverse, and Difference
   of Matrices 190;   4. The Transpose of a Matrix 191;   5. The
   Product of Two Matrices 191;   6. The Identity Matrix, the Inverse
   of a Matrix 193;   7. The Solution of Linear Equations 195;
8-2 MATrix Statements in BASIC 196;
   1. List of MAT Statements 196;   2. DIM, MAT READ, and MAT

PRINT Statements 197; 3. The Zero Matrix, Identity Matrix, Matrix of Ones 201; 4. The Sum, the Difference of Matrices, and Scalar Product Statements 204; 5. The Transpose, Product, Inverse, Determinant Statements 205; 6. Matrix Equality 207; 7. The Same Matrix Variable on Both Sides of an Equal Sign 208; 8. Illegal MAT Statements 210; 9. More about DIM Statements; Vectors 210; 10. N-Dimensional Arrays 212; 11. MAT Statements with Strings 216;

8-3 Solution of Linear Systems Using Matrices 218;
1. Matrix Solution of n Equations with n Variables 218; 2. Program (14): The Solution of Linear Systems Using Matrices 220; 3. Comments about Program (14) 222;

8-4 Summary 224.

CHAPTER 9 225

9-1 A Statistics Package; Program (15) 225;
1. A Statistics Problem 225; 2. Statistical Relationships 225; 3. The Use of Matrices to Tabulate Data 226; 4. The Arithmetic Mean 226; 5. Standard Deviation 227; 6. The Coefficient of Correlation 228; 7. The Percentage of Variance 229; 8. The Equation of the Regression Line 229; 9. Program (15): A Statistics Package 231; 10. Comments about Program (15) 234;

9-2 Area under the Curve of a Function F; Programs (16), (17) 235;
1. Area under the Curve of a Function F 235; 2. The Rectangular Rule 236; 3. The Trapezoidal Rule 238; 4. The Midpoint Rule 240; 5. Simpson's Rule 242; 6. Program (16): Area under the Curve of a Function F 244; 7. Comments about Program (16) 248; 8. Program (17): A Second Approach to the Area under a Curve 249; 9. Comments about Program (17) 254;

9-3 Summary 256.

CHAPTER 10 258

Summary of BASIC Commands 258; I. Special Symbols 258; II. Statements 259; III. Functions 260; IV. String Manipulation 261; V. Operators 262; VI. EDITing Characters 263.

CHAPTER 11 264

11-1 System Commands 264; 11-2 EDIT Commands 267; 11-3 Size Restrictions 269; 11-4 ERROR Messages 270.

APPENDICES: 277

A. Algebra Programs 279
B. Geometry Programs 300
C. Trigonometry Programs 304
D. Analytic Geometry Programs 306
E. Calculus Programs 314
F. Probability Programs 329
G. Special Programs 331

INDEX

Of The More Significant Programs

Program	Page
1. The Area of a Circle and the Volume of a Sphere (1)	62
2. The Area of a Circle and the Volume of a Sphere (2)	69
3. The Area of a Circle and the Volume of a Sphere (3)	78
4. The Real Roots of $AX^2+BX+C=0$	82
5. N! (Factorial)	89
6. N!, A Second Approach	100
7. A Dice Game	108
8. Sum, Product, Maximum, Minimum	114
9. Algebra Quiz	124
10. To Find the Slope of $F(X)$ at $X=C$	144
11. The Solutions of Triangles in Trigonometry	164
12. The Graph of Any Function	174
13. The Real Zeros of Any Function	181
14. The Solution of Linear Systems Using Matrices	220
15. A Statistics Package	231
16. The Area under the Curve of a Function F	244
17. The Area under the Curve of a Function F	250

APPENDICES

Appendix A: Algebra Programs
  Program 1:   Average of N Numbers   279
                  2:    Solution of AX+B=CX+D   279
                  3.   Pairs of Factors of an Integer   280
                  4.   Prime Factors of Any Integer   281
                  5.   Least Common Multiple and Greatest Common Factor   282
                  6.   Solution of Two Simultaneous Linear Equations   284
                  7.   Completion Time of Work Problem   285
                  8.   To Convert a Number from Base 10 to Any Base   286
                  9.   Logarithm of N in Base B   288
                  10.   Approximation of the Square Root of N   289
                  11.   To Find the Rth Root of N   290
                  12.   To Divide One Polynomial by Another   290
                  13.   To Graph a Function with X-Axis Horizontal   292
                  14.   To Graph Up to 26 Functions   293
                  15.   Real Zeros of Any Function by Newton's Method   297

Appendix B: Geometry Programs
                  16.   Pythagorean Triplets   300
                  17.   Area of Any Regular Polygon   301
                  18.   To Find the Interior Angle, Diagonals of a Regular Polygon   302

Appendix C: Trigonometry Programs
                  19.   Area of Triangle, Given SAS or SSS   304

Appendix D: Analytic Geometry Programs
                  20.   Equation of Parabola Through Three Points   306
                  21.   Analysis of $AX^2+BXY+CY^2+DX+EY+F=0$ 308

Appendix E: Calculus Programs
    22. To Find the Limit of Any Function   314
    23. Area Under Curve and Average Value of Any Function   319
    24. Volume of Any Solid of Revolution   322
    25. To Calculate Sin(X) by a Power Series   326

Appendix F: Probability Programs
    26. Binomial Experiment   329

Appendix G: Special Programs
    27. Arranging of Words in Alphabetical Order   331
    28. Electronic Configuration of Any Element   332
    29. Game of Chinese War   334
    30. A Meaningless Technical Report   336

COMPUTER PROGRAMMING IN BASIC

CHAPTER 1

1-1 What Is Time-Sharing?

Until recently, the only way to use a computer was to write a program, punch that program onto cards or tape, take the cards to a central operating area where they would undergo further handling, and finally receive a print-out of the results. The "turn-around" time for such a procedure was an hour or so at best, and possibly a week or longer. Of course, if there was a mistake in the program, the entire procedure had to be repeated after corrections were made.

This procedure for obtaining results on the computer made computer solutions practical only for large problems. The time and frustration discouraged the use of the computer in the solution of small and even medium-size problems.

What was needed was a system which would enable a user to communicate directly with the computer, and by which he could obtain almost instantaneous results from the computer. What was needed was for each of us to have our own computer! Cost seemed to preclude this remedy, however. The next solution to the problem was to develop a system which would permit a number of people to have simultaneous remote access to one large computer. Such a system has been designed. In the system the computer handles a small part of one user's problem, goes on to the next user, handles a part of his problem, etc. In this way, the computer cycles through all of the users who are on the system at that moment. Since the actual computing time for each user is usually quite short, the cycle time is very fast, and each user has the impression that he alone is using the computer. The "hardware" (the actual equipment) and "software" (the programmed instructions) required to carry out this process is called a time-sharing system.

Each user interacts with the computer via a Teletype connected to the computer through conventional telephone lines. Seated at his Teletype, the operator types out his program, feeds it into the computer, and then receives his results typed out on his terminal. This direct interaction between the user and the computer is the single greatest advantage of time-sharing. If there are mistakes in the program, they become immediately evident, and can be corrected on the spot. Thus a procedure which customarily required days or perhaps weeks has been reduced to a matter of minutes or hours.

In addition to the mechanical and electrical hardware required for such a time-sharing system is the software; particularly a language by which the user can communicate with the computer. There are a number of computer languages which one can use to give instructions to a computer. One language which has gained a position of prime importance in the United States today is called **BASIC** (Beginner's All-purpose Symbolic Instructional Code).

1-2 Purpose

The purpose of this book is to teach you how to structure a problem for computer solution and write the necessary program in the BASIC language, and to acquaint you generally with a time-sharing environment. To achieve these goals we shall present a number of BASIC programs in a special sequence. Each program is designed to introduce a particular feature of the BASIC language. The program is then discussed in some detail.

Beginning with Chapter 3, there are exercises at the end of most sections. Read a section, then do the exercises immediately following the section. You should have no difficulty working through this book completely on your own. When you have finished Chapter 11, you will have a good understanding of programming in general, and will be able to write your own programs in BASIC.

CHAPTER 2

2-1 The Teletype

The ASR 33 Teletype is the link between you and the time-sharing system. Examine the diagram of the ASR 33 Teletype in Figure (1), and compare it with the Teletype console itself. (Note that other ASR consoles such as the Teletype Model 35 or Model 37 can be used in the same manner as described in this chapter.)

The ASR 33 is a two-input, two-output device because information can be entered through either the keyboard unit or the tape reader mechanism; information is outputted through the typing unit and the tape punch mechanism. Every signal which can be fed into the computer from the keyboard can also be fed in from the tape reader, but at a much higher rate. Furthermore, every signal received from the computer which can be printed by the typing unit can also be recorded on tape by the punch mechanism.

If you are sending a signal into the computer from the keyboard or from the tape reader, then we say that you are in the ON LINE mode. If the Teletype is not linked with the computer, and you are simply punching a tape, then you are in the LOCAL or OFF LINE mode.

2-2 Punching a Tape

The main purpose of this chapter is to acquaint you with the procedures for (1) punching a tape, (2) linking the Teletype to the computer, (3) feeding the program punched on the tape into the computer or typing a program in directly from the keyboard, and (4) disconnecting the Teletype from the computer. We shall explain these four processes in the order in which you will usually perform them.

To minimize the costs of using a time-sharing system, it is desirable first to punch a tape in the OFF LINE mode, then feed the program punched on the tape into the computer. While it might take you one hour to type out a program at the keyboard, the same program can be fed into the computer through the tape reader mechanism in about five minutes. Since we pay only for ON LINE time, the savings are considerable.

The following procedure should be used when punching a tape OFF LINE.

(1) Depress the LCL button (#18 on the list of Teletype keys in Figure (1).)

(2) Depress the ON button (#4) of the tape punch mechanism.

(3) Strike the HERE IS key on the keyboard once, thus providing sufficient leader on the tape. If holes other than the sprocket holes are punched on the tape, strike the SPACE BAR about ten times, followed by a carriage RETURN, instead of hitting the HERE IS key.

(4) Assume that you have a program you want to transmit to the computer. Begin typing your program. As you type, the typehead will strike the paper and the tape punch mechanism will punch holes in the tape.

(5) When you have completed a line in your program, depress the RETURN key. This action returns the typehead to the beginning of the next line. Later, when you are ON LINE, the holes punched in the tape when you struck the RETURN key will signal to the computer that the line should be stored.

(6) When ON LINE, some systems will automatically line feed the paper after a carriage RETURN. Test your system. If it does not work this way, then depress the LINE FEED key in the OFF LINE mode after each carriage RETURN in order to automatically advance the paper when ON LINE. Needless to say, if the computer automatically line feeds the paper after a carriage RETURN signal, you should advance the paper by hand when punching a tape.

(7) After you have completely typed your program, depress the HERE IS key once (or the SPACE BAR about ten times), thus giving you a "trailer" at the end of your punched tape.

(8) Tear your tape off by a sharp pull upward. Notice that the tear is in the shape of an arrow. Your program begins at the arrow head and ends at the arrow tail.

2-2 Punching a Tape   5

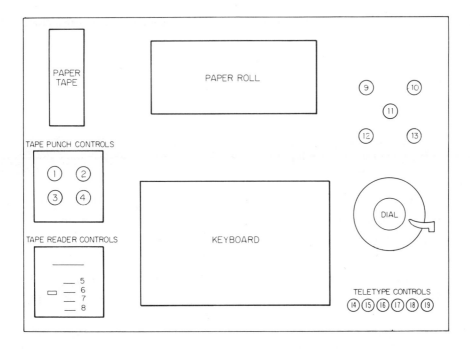

Figure (1). ASR-33 Teletype

TAPE PUNCH CONTROLS

1. RELEASE
2. OFF
3. BACK SPACE
4. ON

TAPE READER CONTROLS

5. MANUAL START
6. AUTO
7. MANUAL STOP
8. FREE

TELETYPE CONTROLS

9. BREAK RELEASE
10. REST
11. FULL DUPLEX
12. OUT OF SERVICE
13. NORMAL/RESTORE

TELETYPE CONTROLS

14. ORIGINATE
15. CLEAR
16. ANSWER
17. TEST
18. LOCAL
19. BUZZER RELEASE

(9) Depress the OFF button (#2) on the tape punch mechanism. You are now ready to feed your program into the computer. But first you must LOG ON, or connect your Teletype to the computer.

2-3 LOGging ON

You are now ready to use the computer. To do so, you must establish a connection, or link, between your Teletype and the computer itself. There is a definite procedure which you must follow.

(1) Press the ORIG button (#14) on the ASR-33 Teletype. The console will turn on and you will hear the characteristic dial tone of a telephone over the built-in speaker. Adjust the volume control on the speaker so that you can hear the dial tone.

(2) Dial your computer time-sharing system telephone number.

(3) You will hear the characteristic ring of a telephone at the other end of the line as the Teletype tries to connect to the computer complex of equipment. When you hear a "beep" you know that the connection with the computer is made.

(4) When the computer asks you for your user number and password, type the appropriate characters.

The LOG ON procedure differs from one system to another. Many computers "echo" what is typed; that is, for each character that you type a signal is sent from the Teletype to the computer, and then from the computer back to the Teletype. This round trip of signals causes a double printing of characters — one from the keyboard and one from the computer. If your system has an echo, the double printing can be eliminated by depressing the FDX button (#11). It will not be apparent to you, but the Teletype will print only the signal returned from the computer. Anything which appears on the paper has actually been received and sent back by the computer.

If you make a mistake in any one of the steps (1) through (4), then the computer will type INVALID USER and will either ask you again for your user number and password, or turn off your Teletype. If your Teletype is turned off, you must dial the number of the computer again. If you cannot LOG ON after several attempts, call the operator of your time-sharing system to determine the status of the system. The operator will not have the same telephone number as the computer.

2-3 LOGging ON

(5) Once you LOG ON, most computers will type:

SYSTEM

The computer is asking you what language you want to use. You type:

BASIC

(6) The computer then types:

NEW OR OLD--

You type:

NEW

and then strike the RETURN key. (The result of typing OLD is described in Section 5-4.) The computer then types:

NEW PROGRAM NAME--

You type a name to identify your program and then hit the RETURN key again. The name of the program is limited to six characters on most systems.

(7) The computer then types:

READY

indicating that it is awaiting its first instruction. You are now ready to feed your taped program into the computer, or to begin typing your program ON LINE.

2-4 Reading In a Punched Tape

To feed your program into the computer from the tape reader, follow these instructions.

(1) Raise the plastic cover on the tape reader by pushing the snap to the right. Insert the tape with the arrowhead facing you and the

small off-center holes positioned on the feeder sprocket. Then snap the plastic cover back into place. See Figure (1) and your console for the location and controls of the tape reader.

(2) Some systems require that you type TAPE and then strike the RETURN key to signal the computer that you want to input information via the tape reader.

(3) Be sure the OFF button (#2) of the tape punch mechanism is depressed, then push the switch of the tape reader to the MANUAL START position (#5). When you release this switch, it will return to its normal position of AUTO (#6). Your tape will begin to feed through the tape reader, and thus your program will be fed into the computer.

(4) The signals from the tape reader go first to an *input buffer* storage area in the computer complex. From there they go to the program or working storage area. As the tape reader sends the signals into the computer, the input buffer storage area (which is quite small) becomes filled with your statements. At this point, the tape reader momentarily pauses. As the input storage area is emptied, the tape reader automatically begins again. Do not restart the tape mechanism during one of these momentary pauses by pushing the tape reader lever switch to the MANUAL START position.

(5) After the program has been completely read into the computer, the machine will pause, awaiting your next command. If you made no mistakes when typing your program, the program is ready to be executed, or RUN. If you wish to make changes in your program, and your computer is one which requires you to type TAPE before feeding in your paper tape, you must now type KEY to signal the computer that you are about to type a message manually. If you are not required to type TAPE, you need not type KEY now, but may continue as if OFF LINE.

We shall not discuss the running of programs now, but shall wait until Section 3-1, at which time you will have written a program that you can actually execute.

2-5 LOGging OFF

After you have executed your program, you are ready to disconnect the Teletype from the computer. The LOGging OFF procedure differs from one time-sharing system to another. It usually consists of typing a set of characters, such as BYE, or LOGOUT which signals the computer to disconnect your Teletype from the computer.

After you type these LOG OFF characters, most computers will type the time, date, and the amount of computer time used, and then automatically turn your Teletype console OFF.

CHAPTER 3

In order for a computer to solve a problem for you, it must have a set of instructions which specify exactly how it should arrive at a solution to that problem. The set of instructions or commands which you give the computer is called a *program*.

The instructions which you give the computer must of course be intelligible to the machine. You must therefore write your commands in a special way, or in a specific language. The language which we shall use in this book is called BASIC (Beginner's All-Purpose Symbolic Instruction Code).

In this chapter we shall introduce you to a number of commands in the BASIC language. We shall also include examples which not only illustrate these commands but also serve as a basis for discussion of the method by which one organizes and writes a program. By studying both the general discussion and the specific examples, you will soon be able to write your own programs in BASIC.

3-1 Command PRINT

One frequently used command is PRINT, followed by whatever information you want the computer to print. The following discussion not only illustrates some of the nuances of the PRINT command but also explains the procedure by which you write a program.

1. Line Numbers

In a BASIC program each line contains one command. Every line begins with a line or step number which must be a nonnegative integer less than 100000. Thus you see in Example (1) that we used the numerals 10, 20, 30, 40 to designate the line numbers.

Recall that at the end of each line you strike the RETURN key. When you are preparing a tape of your program OFF LINE, you usually must line feed the paper manually by rotating the Teletype carriage. When you are feeding in your tape ON LINE however, most computers will automatically line feed the paper after a carriage RETURN. In addition, the carriage RETURN signals the computer that the line has been completed, and instructs the computer to store (remember) that line.

EXAMPLE (1)

PROGRAM:

```
10 PRINT 1 6+7
20P    R I N     T21 +     7
30 PRINT 32    7
40 END
```

OUTPUT:

```
23
28
327
```

2. Spaces

We insert spaces in the statements of our program only for our convenience in reading the program, since **BASIC** ignores all spaces. Thus the computer considers the statement in Step 20 of Example (1) and the statement 20 PRINT 21 + 7 to be equivalent, although the latter is much easier for us to read. Similarly, the statement in Step 30 and the statement 30 PRINT 327 are considered to be equivalent; the computer ignores spaces between all characters, including numbers.

3. Execution of a Program

After you enter your program into the computer (either through the tape reader mechanism or by typing it directly when ON LINE), you type the command RUN, followed by a carriage RETURN. The RUN command tells the computer to execute the stored instructions of the program.

On most systems, before the program is actually executed, the computer types a *heading* usually consisting of the program name, the time and the date. The program name is the name you entered when you first LOGged into the computer in response to NEW PROGRAM NAME. The computer then executes the instructions of the program in numerical order, beginning with the statement with the smallest step number, regardless of the order in which the steps were entered. Thus, if we entered Step 20 of Example (1) after Step 30, the computer would first rearrange the steps in numerical order, placing Step 20 between Steps 10 and 30, and then begin executing the program.

Notice in Example (1) that we did not number our steps consecutively, such as 1, 2, 3 and 4. The reason for this is that, if we must later include a step between, say, Step 10 and Step 20, we can do so by assigning a number between 10 and 20 to the new statement. Numbering our steps consecutively precludes this possibility of later inserting forgotten steps. You will appreciate the real advantage of the nonconsecutive numbering of steps when you begin writing your own programs.

During the execution of Example (1), the computer first evaluates the expression 16 + 7 from Step 10. Because the command given is PRINT, the computer types the value of that expression, namely, 23. The computer automatically returns the carriage and line feeds the paper after it has executed Step 10, and goes on to the next step, 20. The computer evaluates 21 + 7, prints the sum, returns the carriage, and line feeds the paper. The computer continues to execute each step of the program in this fashion.

On most systems, after the computer has finished the program, it types either READY or the time it used the *central processor*. The central processor is that part of the computer which actually carries out the instructions of the program.

In all of the examples given in this book, the heading at the beginning of the OUTPUT and the number of central processor units at the end of the OUTPUT of each program have been omitted.

4. END Statement

On many systems, the statement with the largest step number must be the word END, as in Step 40 of Example (1). Check your system to determine whether or not the word END must be included in your program.

5. Replacing a Line

If two or more statements are preceded by the same line number, then that statement which was entered last is the one which the computer stores in the area associated with that line number. Thus, in

Example (2), we used Step 10 a second time in the program. The new instruction PRINT 32 + 16 replaced the old instruction PRINT 16 + 7, as you can see from the OUTPUT of the program.

EXAMPLE (2)

PROGRAM:

```
10 PRINT 16+7
20 PRINT 21+7
10 PRINT 32+16
30 PRINT 32 7
40 END
```

OUTPUT:

```
48
28
32 7
```

6. Quotation Marks

A *string* is a sequence of letters, digits, or any other characters which are not used to control the system. A string may be printed in the OUTPUT of a program by means of quotation marks. In Step 10 of Example (3) the string is: WHAT APPEARS INSIDE QUOTATION MARKS IS TYPED. During the execution of the program, the computer prints the string in exactly the way we typed the string. The computer does not ignore spaces in a string.

EXAMPLE (3)

PROGRAM:

```
10 PRINT "WHAT APPEARS INSIDE QUOTATION MARKS IS TYPED."
20 END
```

OUTPUT:

```
WHAT APPEARS INSIDE QUOTATION MARKS IS TYPED.
```

7. Commas With Numbers

More than one expression may be written in a single PRINT command by the use of commas. In Example (4), the commas between expressions in Step 15 tell the computer to type the results in columns. The value of 16 + 7 is typed in the first column, 7-18 in the second column, 1/3 in the third column, and so on.

In BASIC there are five columns across the page, each 15 spaces wide. If there are more than five results to be typed, the computer automatically prints the sixth result in the first column of the next line, the seventh result in the second column, etc.

Notice in the OUTPUT of Example (4) that if a number is negative, then a negative sign is printed in the first space of the particular column. If the number is positive, then the first space of that column is left blank. The asterisk * in the expression 16.17*2 is the symbol for multiplication (see Section 3-2).

EXAMPLE (4)

PROGRAM:

```
15 PRINT 16+7, 7-18, 1/3, 2.176+15.382, 21+1.5, 16.17+2, 16.17*2
20 END
```

OUTPUT:

```
 23              -11             .333333          17.558           22.5
 18.17           32.34
```

8. Commas With Strings

Strings may also be printed in a column arrangement. In Example (5), notice that a string inside quotation marks, separated by a comma, produces an OUTPUT with the words ADDITION and SUBTRACTION in the first and second columns. The first letter of each string appears in the first space of the column.

If the OUTPUT of a string in the column arrangement requires more than fifteen characters, as in Step 10 of Example (6), then the next column is utilized to complete the statement. Notice the use of quotation marks enclosing a space in Steps 15 and 17 of Example (6). The quotation marks and space are used so that the OUTPUT of the results of 38.1468-27.4618 and 2.17-6.38 appear in the desired third column, under the heading SUBTRACTION. On some systems, two commas in a row may be used in place of the quotation marks and space.

EXAMPLE (5)

PROGRAM:

```
10 PRINT "ADDITION", "SUBTRACTION"
15 PRINT 3+7.6984, 3.56-4.32
17 PRINT 15.6+17.34, 19.3849-5.6843
20 END
```

OUTPUT:

```
ADDITION        SUBTRACTION
 10.6984         -.76
 32.94           13.7006
```

EXAMPLE (6)

PROGRAM:

```
10 PRINT "MULTIPLICATION AND ADDITION", "SUBTRACTION"
15 PRINT 3*5-7, " ", 38.1468-27.4618
17 PRINT 3*(5-7), " ", 2.17-6.38
20 END
```

OUTPUT:

```
MULTIPLICATION AND ADDITION    SUBTRACTION
  8                             10.685
 -6                             -4.21
```

9. Commas with Numbers and Strings

Step 10 of Example (7) illustrates that numbers and strings may appear in the same PRINT statement. The comma between the string and numerical expression tells the computer to type SUM = in the first column and 10.04, the value of 3.21 + 6.83, in the second column.

Notice the comma at the end of Step 20. During execution, the computer first types the string PRODUCT = . The comma at the end of Step 20 instructs the computer not to print a carriage return and line feed as it usually does when it completes a step, but to continue on the same line to the next column. The computer then executes the next Step 30. Thus you see the number 21.76 on the same line as the string PRODUCT = and in the appropriate second column.

EXAMPLE (7)

PROGRAM:

```
10 PRINT "SUM =", 3.21+6.83
20 PRINT "PRODUCT =",
30 PRINT 3.2*6.8
40 END
```

OUTPUT:

```
SUM =           10.04
PRODUCT =       21.76
```

10. Skipping a Line

The command PRINT alone, as in Step 20 of Example (8), directs the computer to return the carriage and line feed the paper, thus skipping a line in the OUTPUT. You can skip as many lines as you wish using this technique.

EXAMPLE (8)

PROGRAM:

```
10 PRINT "SUM"
20 PRINT
30 PRINT "PRODUCT"
40 END
```

OUTPUT:

```
SUM

PRODUCT
```

11. Semicolon with Numbers

Example (9) illustrates the use of a semicolon to separate expressions in a PRINT statement. The semicolon does not divide the page into five columns of 15 spaces each, as does the comma. Rather, on most systems, the value of the expression is preceded by either a space or a negative sign (dependent on whether the number is positive or negative), and is then followed by a space.

EXAMPLE (9)

PROGRAM:

```
10 PRINT 1, 12, 9-110, -12345, 3+4
20 PRINT 1; 12; 9-110; -12345; 3+4
30 END
```

OUTPUT:

```
1               12              -101            -12345          7
1   12 -101 -12345   7
```

12. Semicolon with Strings

Look at Step 20 of Example (10). Notice that when the semicolon is used to separate strings rather than numbers, no space occurs between the strings in the OUTPUT.

EXAMPLE (10)

PROGRAM:

```
10 PRINT "SUM", "PRODUCT", "ADDITION", "SUBTRACTION"
20 PRINT "SUM"; "PRODUCT"; "ADDITION"; "SUBTRACTION"
30 END
```

OUTPUT:

```
SUM             PRODUCT         ADDITION        SUBTRACTION
SUMPRODUCTADDITIONSUBTRACTION
```

13. PRINT TAB (N) Command

In BASIC, the carriage spaces are labeled 0 through 74, from left to right, thus totaling 75 spaces. The command PRINT TAB (N) instructs the computer to skip to the Nth space on the carriage. In Step 20 of Example (11), PRINT TAB (15) directed the computer to skip 15 spaces to the space numbered 15 on the carriage. The command PRINT TAB (42) in Step 30 directed the computer to skip to the 42nd space on the carriage; the first three spaces are occupied with the word SUM, followed by 39 blank spaces.

Step 40 shows that the computer ignores a PRINT TAB (N) command if it has already passed the Nth space when it reads that command.

The argument N of the PRINT TAB (N) command may be any nonnegative number or numerical expression. If $75 \leq N < 150$, then TAB (N) = TAB (N-75); if $150 \leq N < 225$, then TAB (N) = TAB(N-2*75), etc. We say that TAB (N) is modulo 75. Thus in Step 50, TAB (90) caused the computer to skip the first 15 spaces, since TAB (90) = TAB (90-75) = TAB (15).

In Step 60, the computer first evaluated the argument (11 + 4.996) = 15.996, ignored the decimal portion of this argument, and used TAB (15) to skip the first 15 spaces on the line. Whenever the argument N of the PRINT TAB (N) command is not an integer, the computer ignores the decimal portion of that argument.

EXAMPLE (11)

PROGRAM:

```
10 PRINT "SUM", "PRODUCT"
20 PRINT TAB(15); "PRODUCT"
30 PRINT "SUM"; TAB(42); "ADDITION"
40 PRINT "SUM"; "PRODUCT"; TAB(5); "ADDITION"
50 PRINT TAB(90); -6.5; 17
60 PRINT TAB(11+4.996); -3.6
70 END
```

OUTPUT:

```
SUM             PRODUCT
                PRODUCT
SUM                                     ADDITION
SUMPRODUCTADDITION
               -6.5   17
               -3.6
```

14. Legal Expressions in PRINT Statements

The expression to the right of the command PRINT must be one or more of the following:

    (1) number
    (2) numerical expression
    (3) string (enclosed in quotation marks)
    (4) numerical variable (see Section 3-4)
    (5) string variable (see Section 5-4)
    (6) function (see Section 4-1 and Section 6-4).

Step 10 of Example (12) shows three numerical variables B, C3 and X(5) in a PRINT statement, and illustrates the fact that the computer assumes that all variables are equal to zero unless they have been defined in the program. A complete discussion of variables will be found in Sections 3-4 and 5-4.

Step 20 illustrates the use of a function which the computer understands, namely, the square root function. Notice in Step 30 that the computer typed what was inside the quotation marks; it did not evaluate SQR(625).

EXAMPLE (12)

PROGRAM:

```
10 PRINT B, C3, X(5)
20 PRINT SQR(625)
30 PRINT "SQR(625)"
40 END
```

OUTPUT:

```
0               0               0
25
SQR(625)
```

15. Exponential Notation

Notice in Example (13) that we typed our numbers in exponential notation. The general rule for interpreting this notation is as follows:

$$bEp = b \times 10^p,$$

where b is either an integer or a decimal numeral, and p is an integer. Thus, in Step 10 of Example (13),

$$1.52E3 = 1.52 \times 10^3 = 1520$$

Notice in Step 30 that we wrote 1E5 to represent the number 100000. You cannot omit the 1 before the E.

EXAMPLE (13)

PROGRAM:

```
10 PRINT 1.52E3, 15.2E2, 0.152E4
20 PRINT 3.67E-2, 367E-4
30 PRINT -85.1E2, 1E5
40 END
```

OUTPUT:

```
 1520          1520          1520
 .0367         .0367
-8510          100000
```

The computer can also type numbers in exponential notation, as you can see in the OUTPUT of Example (14). $-1.70141 \text{ E} + 38$ means -1.70141×10^{38}. Although a complete discussion of numbers in BASIC will be presented in Section 4-2, you should now realize that when a number appears in exponential notation in the OUTPUT, the computer types the following:

(1) either a space or a negative sign, depending on whether the number is positive or negative
(2) the first digit of the number
(3) a decimal point
(4) the remaining significant digits of the number
(5) a space
(6) the letter E
(7) the sign + or -
(8) the appropriate exponent number.

EXAMPLE (14)

PROGRAM:

```
10 PRINT LOG(A)
20 END
```

OUTPUT:

```
LOG OF ZERO IN 10
-1.70141 E+38
```

16. ERROR Messages

When the computer does not understand a given statement, or when there is an unusual condition in your program, the machine generally replies with what we call an ERROR message. Example (14) illustrates the typing of an ERROR message. In Step 10, the computer first substitutes 0 for the value of A. (Recall that the computer assumes that the value of a variable is 0 if it has not been previously defined in a program.) It then attempted to calculate LOG(0), using the built-in natural logarithmic function (see Section 4-1). Since LOG(0) is not a well-defined number, the computer typed an appropriate ERROR message LOG OF ZERO IN 10.

After typing the ERROR message, the computer attempted to evaluate LOG(0) by calculating LOG(N), where N is very close to 0. The value of LOG(0) was given as $-1.70141\ E+38$, the largest value in a negative sense on the system used in this example. If you have studied the logarithmic function, you know that as N approaches 0 from the right, then LOG(N) approaches negative infinity.

EXERCISES 3-1

Punch a tape of the statements (1) through (7) OFF LINE, and strike twice the HERE IS key between each problem. LOG ON to the computer. (Refer to Section 2-3 if necessary.) Feed Problem (1) only through the tape reader, and then push the lever of the tape reader mechanism to the MANUAL STOP (#7 on the diagram in Section 2-2) position. Execute Problem (1) by typing RUN ®. After Problem (1) has been executed, push the lever to the MANUAL START (#5) position and feed Problem (2) through the reader. Continue this process until all problems have been executed. Note that ⓡ means that you hit the RETURN key.

1. 10 PRINT 3.15-6.3741, 5.7+3.64, 316, 0.21-7.3, 17 ⓡ
   20 END ⓡ

2. 10 PRINT A ⓡ

3. 10 PRINT LOG (3.14159) ⓡ

4. 15 PRINT THIS IS EASY ⓡ

5. 10 PRINT "ONE, TWO, THREE, FOUR, FIVE, SIX" ⓡ
   15 PRINT ⓡ
   17 PRINT "ONE", "TWO", "THREE", "FOUR", "FIVE", "SIX" ⓡ

6. 10 PRINT 324, -1963.5, 21.6 ⓡ
   13 PRINT 325; -1963.5; 21.6 ⓡ
   15 PRINT "ONE; TWO; THREE; FOUR" ⓡ
   17 PRINT "ONE"; "TWO"; "THREE"; "FOUR" ⓡ

7. 10 PRINT 5+5; 10+10; 2-3; 100+999.6 ⓡ
   13 PRINT TAB (4); 10+10; TAB(11); 100+999.6 ⓡ
   15 PRINT 2/3, 10.3712+6.3374; TAB (10); 3-2 ⓡ
   17 PRINT "ONE"; TAB (4); "TWO" ⓡ
   19 PRINT TAB (85.9); "FIRST" ⓡ

3-2 Operations ↑, *, /, +, -

1. Symbols Used for Arithmetic Operations

In BASIC, the following symbols are used to denote arithmetic operations:

Symbol	Operation
↑	Exponentiation
*	Multiplication
/	Division
+	Addition
-	Subtraction

2. Order of Operations

When evaluating a given expression, the computer performs the above operations in a specific order:

(1) all exponentiations first
(2) all products and quotients second, in the order in which they occur, from left to right
(3) all sums and differences third, from left to right.

3. Use of Parentheses

You can change the above order of operations by the use of parentheses. The computer evaluates the expression in parentheses first, then continues the evaluation according to the above rules.

For example, the computer would evaluate the expression 4∗2 ↑ 3 + 6/3 as follows:

$$4*2 \uparrow 3+6/3 \ = 4*8+6/3$$

$$= 32+2$$

$$= 34$$

However, if the original expression contained parentheses, such as 4∗2 ↑ (3 + 6/3), the computer would react as follows:

$$4*2 \uparrow (3+6/3) = 4*2 \uparrow (3+2)$$

$$= 4*2 \uparrow 5$$

$$= 4*32$$

$$= 128$$

Parentheses may be contained within other parentheses, such as 4∗2 ↑ ((3+6)/3). The computer evaluates the innermost expression first, and works outward:

$$4*2 \uparrow ((3+6)/3) = 4*2 \uparrow (9/3)$$

$$= 4*2 \uparrow 3$$

$$= 4*8$$

$$= 32$$

The following Example (1) on the computer illustrates the rules of order and the use of parentheses. Study each expression carefully so that you understand completely the order of precedence of the operations of arithmetic and the use of parentheses to change this order.

EXAMPLE (1)

PROGRAM:

```
 5 PRINT 2+6/2, 2+6/2-4*3, 2+6/2-4*3↑2, (2+6)/2, (2+6)/(2-4)*3
10 PRINT (2+6)/2-(4*3)↑2, 2+(6/2-4)*3↑2, 2+(6/(2-4))*3↑2
15 PRINT (2+(6/2-4)*3)↑2, (2+(6/(2-4)*3))↑2, ((2+(6/(2-4)*3))↑2)/7
20 PRINT (2+(6/(2-4)*3))↑2/7
25 END
```

OUTPUT:

```
  5              -7          -31              4           -12
-140             -7          -25
  1              49            7
  7
```

EXERCISES 3-2

To do the following problems, follow the directions given for Exercises 3-1. In Problem (2), write each expression on one line on a separate sheet of paper before punching your tape.

1. 10 PRINT 315; 5*12; 67-23; 3↑4; 2↑(1/2)
   20 PRINT 4+3↑2; (4+3)↑2; 4*3↑2; (4*3)↑2
   30 PRINT 3*9/4↑2; (3*9)/4↑2; 3*(9/4)↑2
   40 PRINT 3*9/4↑(-2); 3*(9/4)↑(-1/2); (-32)↑(1/5)
   50 PRINT 3*4+2*3; 3*(4+2)*3; 3*(4+2*3)
   60 PRINT 3↑1/2; (3)↑(1/2); -(3)↑(1/2)
   70 PRINT (1776)↑(2/3); (37.4689)↑(2/3); 3*(9/4)↑(1/2)
   80 PRINT (9+5.0713)/(-16.3)*17↑.5
   90 END

2. a) $\dfrac{6(2^3)}{3(-2)}$

   b) $\left(\sqrt{\dfrac{32}{5+3}}\right)^3$

   c) $47.21 - \sqrt[3]{-27} - 16^{3/4}$

d) $1 + \dfrac{1}{2 + \dfrac{1}{2 + \dfrac{1}{2}}}$

e) $\dfrac{\sqrt{3.9}\ (7.4 - \sqrt[3]{8.79})}{-0.5}$

3-3 Corrections

Suppose, when you are typing OFF LINE, you typed the command:

10 PEINT 2+3

You misspelled a word. When you feed the tape into the computer and type RUN, the machine will give you an ERROR message of ILLEGAL INSTRUCTION IN LINE 10, indicating that it does not understand that command. (You must realize that the computer does not "know" the word PRINT; rather, it accepts a configuration of electronic signals associated with the word PRINT. When you type PEINT you trigger a different arrangement of signals which the computer has not been previously programmed to accept. Under these conditions, the computer has been programmed to type the ERROR message.)

You would, however, like to be able to correct your misspelled word PEINT and any other mistakes you make. We shall now discuss several methods of making corrections.

1. SHIFT O Key (Back Arrow)

By depressing one of the SHIFT keys and holding it down while you also depress the O key, you will delete the last character typed. You can delete as many previously typed characters as you like by striking the O key, one strike for each character, including a space. The symbol ← will be typed for each SHIFT O.

Example (1) shows the use of the SHIFT O key to make corrections.

EXAMPLE (1)

PROGRAM:

```
10 PEINT←←←RINT 2+3
40 END
```

OUTPUT:

```
5
```

Another example of making corrections using the SHIFT O key is as follows.

EXAMPLE (2)

PROGRAM:

```
10 PE←RINT 7+12
20 PRINT "THIS ERRE←OR OCCURS INSIDE QUOTATION MARKS."
30 PRINT 17.001-16.1936←←←001
40 END
```

OUTPUT:

```
19
THIS ERROR OCCURS INSIDE QUOTATION MARKS.
 1
```

2. Retyping a Line

If after typing a line you notice that you made a mistake, you may correct that mistake by simply retyping the line using the same step number. The computer stores that statement which was typed last for a given line number.

Even if you have not finished typing a line and notice a mistake at the beginning of the line, you may use this same technique. You simply strike the RETURN key and retype the line correctly, using the same line number. The following two examples illustrate the use of the RETURN key to correct a line.

EXAMPLE (3)

PROGRAM:

```
10 PRINT 2+3
10 PRINT 2+16
20 END
```

OUTPUT:

```
18
```

EXAMPLE (4)

PROGRAM:

```
10 PRINT 2+3
10 PPRINT 2+
10 PRINT 2+18
20 END
```

OUTPUT:

```
20
```

3. ALTMODE (ESCAPE or PREFIX) Key

Hitting the ALTMODE key (on Teletype Model 35) or ESCAPE key (on Model 33) or PREFIX key (on Model 37) deletes the entire line you are typing. The computer prints $--DELETED or simply DELETED, followed by a carriage RETURN and LINE FEED. The computer will ignore that line containing the ALTMODE signal.

Hence, you can make corrections OFF LINE by hitting the ALTMODE key once. The appropriate holes will be punched on the tape, but no character will be typed on the paper. You can then go on to the next step, or retype the same step correctly, by:

(1) turning the tape punch mechanism OFF
(2) hitting the RETURN and LINE FEED keys
(3) turning the tape punch back ON.

The following example illustrates this procedure while OFF LINE. Ⓔ represents a stroke on the ESCAPE key.

EXAMPLE (5a)

PROGRAM:

```
10 PRINT 3+34.45
10 PRINT "DELETE THIS LINE BY HITTING ESCAPE KEY ONCE"  (E)
20 END
```

This is what you see as you feed the OFF LINE tape through the reader when ON LINE, and then execute the program.

EXAMPLE (5b)

PROGRAM:

```
10 PRINT 3+34.45
10 PRINT "DELETE THIS LINE BY HITTING ESCAPE KEY ONCE"$--DELETED
20 END
```

OUTPUT:

```
37.45
```

Note that the ALTMODE key cannot be used to delete a line after you have struck the carriage RETURN key to terminate that statement. You can easily delete the statement, however, by typing its step number followed by a carriage RETURN.

EXERCISES 3-3

Type the following problem OFF LINE, LOG ON to the computer, feed the tape through the reader, RUN the program, then LOG OUT.

```
1. 10 PRINT "THIS SENTENCE IS CORRECT"
   10 PRINT "THERE IS AN ERROR    (E)
   20 PRINT "THIA SENTENCE HAS AN   (R)
   20 PRINT "THIS SENTENCE HSS←←AS AN ERROR."
   30 END
```

3-4 Command LET

In BASIC there are three important methods of assigning values to variables:

(1) LET statements
(2) READ-DATA statements
(3) INPUT statements.

In this section we shall consider the first method: LET statements.

1. Assignment of Values to Variables

In Steps 10 and 20 of Example (1) we assigned values to the variables A and B. The computer assigns the number 2 to a particular storage area, and the number 3 to another storage area. In Step 30 the computer stores in a third storage area the sum of the values which it previously stored in the two storage areas which *we* called A and B. In Step 40 the machine then types the value of the variable which it stored in area C, namely, 5.

In Step 30 the statement LET C = A+B does not represent an equation. The LET command instructs the computer first to evaluate the expression on the right side of the equal sign and then to store the result as the value of the variable on the left of the equal sign.

EXAMPLE (1)

PROGRAM:

```
10 LET A=2
20 LET B=3
30 LET C=A+B
40 PRINT C
50 END
```

OUTPUT

5

In Step 10 of Example (2) we assigned the value 2 to variable A. Then in Step 40 we assigned a new value to the *same* variable A, so that in Step 50 we might expect the value of C to be 7. However, since the computer executes a program in numerical order and since the value of C was assigned in Step 30, the number in storage area C is still 5, as you can see in the OUTPUT.

EXAMPLE (2)

PROGRAM:

```
10 LET A=2
20 LET B=3
30 LET C=A+B
40 LET A=4
50 PRINT C
60 END
```

OUTPUT:

5

If we include another assignment statement for the variable C *after* Step 40, as in Example (3), then the OUTPUT for C will change accordingly.

EXAMPLE (3)

PROGRAM:

```
10 LET A=2
20 LET B=3
30 LET C=A+B
40 LET A=4
50 LET C=A+B
60 PRINT C
70 END
```

OUTPUT:

7

Thus in Example (3) we assigned a *new* value to C in Step 50, namely, 4+3 = 7. Consequently the value of C in the OUTPUT is 7.

2. Legal Expressions in LET Statements

The expression on the right side of the equal sign in a LET statement may consist of any combination of constants, variables, and functions, joined by any of the five operators of arithmetic. The computer evaluates the expression on the right side of the equal sign and assigns that value to the variable on the left side of the equal sign.

Example (4) illustrates one combination of legal expressions in the LET statement. Study Example (4) carefully so that you understand exactly how the computer determined the number 30000 as the value for Y.

EXAMPLE (4)

PROGRAM:

```
10 LET X=2↑3
20 LET Y=(COS(0)+X)/3*1E4
30 PRINT X,Y
40 END
```

OUTPUT:

```
 8              30000
```

Example (5) illustrates that the computer will not accept an expression to the left of the equal sign.

EXAMPLE (5)

PROGRAM:

```
10 LET D+1=E
20 END
```

OUTPUT:

```
ILLEGAL FORMAT IN 10
```

3. Legal Names for Variables

In BASIC a variable may be a *single* letter or a *single* letter followed by a *single* digit, as in Example (6).

EXAMPLE (6)

PROGRAM:

```
10 LET D=5
20 LET A1=6
30 LET A(1)=10
40 LET A(D)=11
50 PRINT D, A1, A(1), A(5), Q
60 END
```

OUTPUT:

```
5         6         10        11        0
```

Example (7) illustrates: in Step 10 that the computer does not accept a letter followed by two digits; in Step 20 that we cannot write the digit first and then the letter; and in Step 30 that the machine will not accept two letters for variables.

EXAMPLE (7)

PROGRAM:

```
10 LET A11=7
20 LET 1A=8
30 LET AA=9
40 PRINT A11, 1A, AA
50 END
```

OUTPUT:

```
ILLEGAL FORMAT IN 10
ILLEGAL FORMAT IN 20
ILLEGAL FORMAT IN 30
ILLEGAL FORMAT IN 40
```

Steps 30 and 40 of Example (6) illustrate the correct use of *subscripted* variables. Since the Teletype types only one line at a time, we write the subscripted variable A_1 as A(1). (A complete discussion of subscripted variables will be given in Section 5-8). In Step 40, since the value of D is 5, the computer sets A(5) equal to 11. Thus in Step 50 the machine prints 11 for the value of A(5). You may use any variable or numerical expression as a subscript. In Step 50 of Example (6), since Q was not defined, the value for Q which is given in the OUTPUT is 0.

4. The Same Variable on Both Sides of an Equal Sign

One of the most useful applications of the LET command is illustrated in Example (8). The statement LET D = D+1 in Step 20 enables us to increment the value of D by one. That is, the previously stored value of D was 5, from Step 10. The computer evaluates the expression to the right of the equal sign by adding 1 to this value of D, and then stores the new value 6 in the same storage area D. Of course, the old value of D, namely, 5, is erased when the new value of D is stored.

EXAMPLE (8)

PROGRAM:

```
10 LET D=5
20 LET D=D+1
30 PRINT D
40 END
```

OUTPUT:

```
6
```

5. The Assignment of the Same Value to Several Variables

On most systems a single LET statement may be used to assign the same value to several variables. Step 10 of Example (9) assigns the value 9 to the three variables A, C2, and F(5), as you can see in the OUTPUT.

EXAMPLE (9)

PROGRAM:

```
10 LET A=C2=F(5)=3↑2
20 PRINT A, C2, F(5), 3↑2
30 END
```

OUTPUT:

```
9          9          9          9
```

6. The Word LET Omitted

On some systems you can omit the word LET in the assignment of variables. The computer still understands what you mean, as shown in Example (10).

EXAMPLE (10)

PROGRAM:

```
10 D=6
20 D=D+3
30 PRINT D
40 END
```

OUTPUT:

```
 9
```

EXERCISES 3-4

Follow the directions given in Section 3-1 for Problems (1) through (5) below.

1.
```
10  LET A = 1
20  LET B = A+4
30  LET C = 2*B+A+5
40  PRINT A, B, C
50  END
```

2.
```
40  LET D(1) = 3
50  LET D(2) = 4
60  LET D(3) = 5
70  PRINT D, D(D(1))
80  END
```

3.
```
70   LET E = 1
80   LET E = E+1
90   PRINT E
100  END
```

4.
```
90   LET F = 2*E+D(2)-C
100  PRINT F
110  END
```

5. 100 F = F+10
   110 G = Q(7) = A6 = 4 ↑ 2+1
   120 PRINT F, G, Q(7), A6
   130 END

3-5 READ and DATA Statements

In BASIC a second method of assigning values to variables is by using READ and DATA statements. The two are used together in a program.

1. Method of Assigning Values

As the computer executes the program in Example (1), it first READs A in Step 10. The first READ command encountered in a program instructs the computer to find the first DATA statement. Hence the computer searches through the program, in numerical order, until it finds in Step 30 the statement DATA 13. The computer then assigns the value 13 to the variable A. The computer continues to execute the program by printing the value of A in Step 20.

EXAMPLE (1)

PROGRAM:

```
10 READ A
20 PRINT A
30 DATA 13
40 END
```

OUTPUT:

13

2. The Order of Assignment of Values

The values of the variables are assigned in the order in which they appear in the DATA statements. Hence in Example (2) the value 4 was assigned to A, since A is the first variable in Step 40, and 4 is the first item of DATA in the program.

The DATA may be entered in any number of DATA statements and may appear anywhere in the program. Most programmers write their DATA at the end of the program.

36   Computer Programming in BASIC

Notice in Step 20 of Example (2) that variables in READ statements are separated by commas. The numbers in the DATA statements are also separated by commas where necessary, as you can see in Step 30.

EXAMPLE (2)

PROGRAM:

```
10 DATA 4
20 READ A,B,C,D
30 DATA 3,1,2
40 PRINT A,B,C,D
50 END
```

OUTPUT:

4	3	1	2

Example (3) shows how the computer executes a program containing more than one READ statement. Step 10 instructs the computer to READ the first two values in the DATA statement of Step 50 and to place them in storage areas A5 and P(2) respectively. Step 20 directs the computer to type these values. Step 30 causes the computer to READ three more values from the DATA. The computer begins reading the DATA from where it left off in the last READ statement. Thus the computer sets X = 9, Y = 6, and Z = 11.

EXAMPLE (3)

PROGRAM:

```
10 READ A5,P(2)
20 PRINT A5,P(2)
30 READ X,Y,Z
40 PRINT X,Y,Z
50 DATA 2,4,9,6,11
60 END
```

OUTPUT:

2	4	
9	6	11

3. The Placement of READ Statements

The READ statements must appear before the variables are used in the program; otherwise the variables will be undefined. Hence in Example (4) the OUTPUT gave 0 for the value of A, and not 5 as intended, because the computer was instructed to PRINT A in Step 10 before it read the value of A.

EXAMPLE (4)

PROGRAM:

```
10 PRINT A
20 READ A
30 DATA 5
40 END
```

OUTPUT:

```
0
```

4. Legal Numbers in DATA Statements

DATA statements may contain only numbers. The ERROR message in Example (5) indicates that the computer does not accept expressions such as 2+3 and 5*2 in DATA statements.

EXAMPLE (5)

PROGRAM:

```
10 READ A,B,C
20 PRINT A,B,C
30 DATA 2+3,5*2,6-4
40 END
```

OUTPUT:

```
ILLEGAL FORMAT IN 30
```

However, the numbers may be written in exponential form, as in Step 10 of Example (6).

EXAMPLE (6)

PROGRAM:

```
10 DATA 3E5,4E2
20 READ A,B
30 PRINT A,B
40 END
```

OUTPUT:

```
300000          400
```

5. Insufficient DATA

If there is insufficient DATA for the number of variables to be read, the computer, during execution, gives the appropriate ERROR message and stops executing the program, as in Example (7).

EXAMPLE (7)

PROGRAM:

```
10 READ A1,A2,A3,A4
20 DATA 31,12,5
30 PRINT A1,A2,A3,A4
40 END
```

OUTPUT:

```
OUT OF DATA IN 10
```

However, if you have more DATA than variables, the computer simply ignores those values which it does not use, as in Example (8).

6. The Command RESTORE

The command RESTORE in a program causes the DATA *pointer* to go back to the beginning of the DATA list.

The OUTPUT of Example (9) indicates that the variables A, B, C, D, E, F were assigned the values 1, 2, 3, 4, 5, 6 respectively.

EXAMPLE (8)

PROGRAM:

```
10 READ B1,B2,B3
20 PRINT B3;B2;B1
30 DATA 10,20,30,40,50
40 END
```

OUTPUT:

```
30   20   10
```

EXAMPLE (9)

PROGRAM:

```
10 READ A,B,C
20 READ D,E,F
30 PRINT A,B,C
40 PRINT D,E,F
50 DATA 1,2,3,4,5,6
60 END
```

OUTPUT:

```
1             2             3
4             5             6
```

In Example (10) we inserted one step in the program of Example (9): 15 RESTORE. In Step 10 of Example (10), the computer assigned the values 1, 2, 3 to A, B, C respectively. The command RESTORE in Step 15 caused the DATA pointer to go back to the beginning of the DATA list. Then in Step 20, the values 1, 2, 3 were assigned to the variables D, E, F respectively. The numbers 4, 5, and 6 in the DATA list were ignored by the computer.

EXAMPLE (10)

PROGRAM:

```
10 READ A,B,C
15 RESTORE
20 READ D,E,F
30 PRINT A,B,C
40 PRINT D,E,F
50 DATA 1,2,3,4,5,6
60 END
```

OUTPUT:

1	2	3
1	2	3

EXERCISES 3-5

Follow the directions given in Section 3-1 for Problems (1) through (7) below.

1. ```
   10 READ A1, A2, A3
   20 DATA 5, 2.369, 7.4
   30 PRINT A3, A1, A2
   40 END
   ```

2. ```
   10 READ C
   20 DATA 5↑6
   30 PRINT C
   40 END
   ```

3. ```
   10 READ Z(1), Z(3), Z(2), Z(4)
   20 PRINT Z(1), Z(2), Z(3), Z(4)
   30 DATA 10, 12
   40 DATA 11
   50 DATA 13
   60 END
   ```

4. ```
   10 READ A1, A2, A(1), A(2)
   20 READ B
   30 READ C, D
   40 DATA 1,2,3,4,5,6,7
   50 PRINT A1, A2, A(1), A(2), B, C, D
   60 END
   ```

5. 25 RESTORE

6. 10 READ A
   20 READ B
   25 READ C
   30 READ D
   40 DATA 1, 5, 7
   50 PRINT A, B, C, D
   60 END

7. 10 DATA 5,7,9, 1.6, 2.35
   20 PRINT A1, A2, A3, A4, A5
   25 READ A1, A2, A3, A4, A5, A6
   30 DATA 17.656,2,1
   40 PRINT A1, A2, A3
   50 PRINT A4, A5, A6
   60 END

3-6 The INPUT Command

The third method of assigning values to variables is the command INPUT. This command can be used during the execution of a program, and thus provides for interaction between you and the computer.

1. Using the INPUT Command

As the computer executes Step 10 of Example (1), it types a question mark and waits for you to type in a value for Q. After you type your value of Q and strike the RETURN key, the computer continues to execute the program. In Example (1) we typed the number 17 for Q. Thus, the OUTPUT is Q = 17 from Step 20.

Notice the format of the PRINT statement in Step 20. During the execution of Step 20, the computer first types the string Q = , and then types the value of Q. In a PRINT statement, if there is no punctuation mark between a string enclosed in quotation marks and a variable, then the computer prints the value of the variable immediately after the string. The space to the right of Q = in the OUTPUT is due to the fact that 17 is a positive number.

2. INPUTting Several Variables

More than one variable may be entered in one INPUT statement, as in Step 20 of Example (2). Notice that the variables in the INPUT statement are separated by commas.

EXAMPLE (1)

PROGRAM:

```
10 INPUT Q
20 PRINT "Q =" Q
30 END
```

OUTPUT:

```
? 17
Q = 17
```

When executing the program, the computer first types the string from Step 10. The semicolon at the end of the string suppresses the carriage return and line feed. The INPUT statement in Step 20 directs the computer to type a question mark and to pause for you to type in your values of A1 and A2. Notice that the question mark is typed immediately after the string. You type your values of A1 and A2, separated by a comma, and hit the RETURN key. The computer then continues to execute the program.

You may enter as many variables as will fit on one line in an INPUT statement.

Notice the technique of printing the OUTPUT of Step 40. The computer types a space (because A1 is positive), the value of A1 (13), a space (because of the semicolon), the "+" sign, a space (because A2 is positive), the value of A2 (28), a space (because of the semicolon), the "=" sign, a space (because A1 + A2 is positive), and the value of A1+A2 (41).

EXAMPLE (2)

PROGRAM:

```
10 PRINT "WHAT TWO NUMBERS DO YOU WANT TO ADD";
20 INPUT A1,A2
30 LET A=A1+A2
40 PRINT A1;"+"A2;"="A
50 END
```

OUTPUT:

```
WHAT TWO NUMBERS DO YOU WANT TO ADD? 13,28
 13 + 28 = 41
```

3. Legal Numbers in INPUT Statements

Numbers in exponential notation are accepted in INPUT statements, but expressions such as 3+2 and 4↑5 are not, as you can see in Example (3). Notice in Step 17 that the computer gave us an ERROR message and asked us to retype our INPUT data.

EXAMPLE (3)

PROGRAM:

```
10 INPUT P
15 PRINT P
17 INPUT Q,R
20 PRINT Q,R
30 END
```

OUTPUT:

```
? 5E2
  500
? 3+2, 4↑5
INPUT DATA NOT IN CORRECT FORMAT--RETYPE IT
? -3,9.52
 -3            9.52
```

EXERCISES 3-6

Follow the directions given in Section 3-1 for Problems (1) and (2) below.

1. ```
   10 INPUT A(1)
   20 PRINT A(1)
   30 END
   ```

2. ```
   10 PRINT "WHAT IS THE EXPONENT";
   20 INPUT A
   30 PRINT "WHAT IS THE NUMBER TO BE RAISED TO THE" A;
   40 INPUT B
   50 PRINT B; " ↑ " A; "=" B ↑ A
   60 END
   ```

3-7 Summary

In this chapter you have learned how to set up and write simple programs in the BASIC language, using the commands PRINT, LET, READ and DATA, and INPUT.

The PRINT command is an OUTPUT statement which instructs the computer to type values of expressions, functions, and variables. The LET, READ and DATA, and INPUT commands are methods by which you can store and operate on values.

You have learned about the order of the five operations of arithmetic ($\uparrow$, $*$, $/$, $+$, $-$), and how parentheses can be used to alter this order.

You have also learned about exponential notation, and that the computer will always accept numbers written in this form.

CHAPTER 4

The BASIC language has been designed primarily as a language for computation. In addition to the computer's ability to perform the five fundamental operations of arithmetic (↑, *, /, +, -), it can also work with certain mathematical functions, such as square root, logarithm, sine, and tangent. In this chapter you will learn how to use these and other functions on the computer. You will also learn more about the kinds of numbers used in BASIC.

4-1 Built-In Functions

A number of mathematical functions have been incorporated into the BASIC language to help you, when necessary, with your computations. Before we discuss these functions, we shall give you a brief background of what a function is. You can learn more about functions by studying any modern mathematics textbook on the subject.

1. What Is a Function?

Recall from your work in mathematics that a function is a set of ordered pairs (x, f(x)) such that for a given value of x there exists a unique value f(x). A function may also be thought of as a correspondence between two sets A and B which maps an element x from set A into a unique element f(x) in set B. We sometimes call x the *argument* of the function, and f(x) the *value* of the function. The set A is called the *domain,* and the set B is called the *range* of the function f. Some mathematicians call the set B the *image* set, and set A the *pre-image* set.

For example, we can define a function f by the rule $f(x) = x^2$. To each real number x, there is associated a non-negative real number f(x) which is obtained by squaring the number x. In this example, if the domain A is the set of all real numbers, then the range B is the set of all non-negative real numbers. Some elements which belong to this particular function f are:

$$\{ (0,0), (1,1), (-1,1), (3,9), (-6,36), (\sqrt{2},2) \}.$$

2. Functions in BASIC

The following functions are included in the BASIC language.

Function	Description
ABS(X)	Absolute value of X ($|X|$)
SGN(X)	Sign of X (+1 if $X > 0$, 0 if $X = 0$, −1 if $X < 0$)
INT(X)	Greatest integer less than or equal to X ($[X]$)
RND(X)[1]	Random number between 0 and 1
SQR(X)	Square root of $|X|$ ($\sqrt{|X|}$)
LOG(X)	Natural logarithm of $|X|$ ($\ln|X|$)
EXP(X)	Exponential in base e (e^X)
SIN(X)	Sine of X (X is in radian measure)
COS(X)	Cosine of X (X is in radian measure)
TAN(X)	Tangent of X (X is in radian measure)
COT(X)[2]	Cotangent of X (X is in radian measure)
ATN(X)	Arctangent of X (result is in radian measure)

[1] The argument of the RND function may be omitted on some systems.
[2] The cotangent function is not included on some systems.

The argument X of these functions may be a well-defined number, a variable, another function, or an expression. When evaluating a function, the computer first evaluates the argument if necessary, and then finds the value of the function. Notice that the argument X must be enclosed in parentheses.

3. Absolute Value and Sign Functions

The absolute value function, written ABS in BASIC, is defined as follows:

$$ABS(X) = \begin{cases} X & \text{if } X \geq 0 \\ -X & \text{if } X < 0 \end{cases}$$

Thus the OUTPUT of Step 10 in Example (1) gives 3.2 for the value of ABS(-3.2), 16.5 for ABS(16.5), and 0 for ABS(0).

EXAMPLE (1)

PROGRAM:

```
10 PRINT ABS(-3.2), ABS(16.5), ABS(0)
20 PRINT SGN(-5.634), SGN(0), SGN(5.983)
30 END
```

OUTPUT:

```
 3.2           16.5            0
- 1             0              1
```

The function SGN is the sign of its argument X:

$$SGN(X) = \begin{cases} +1 & \text{if } X > 0 \\ 0 & \text{if } X = 0 \\ -1 & \text{if } X < 0 \end{cases}$$

as Step 20 of Example (1) illustrates.

4. The Greatest Integer Function; Rounding Off

The function INT is the greatest integer function. That is, INT(X) is the greatest integer J which is less than or equal to X. Thus in Example (2) INT(4.28534) = 4, whereas INT(-4.28534) = -5, since -5 is the greatest integer which is less than or equal to -4.28534. The greatest integer which is less than or equal to -4 is -4.

EXAMPLE (2)

PROGRAM:

```
10 PRINT INT(4.28534), INT(-4.28534), INT(-4)
20 END
```

OUTPUT:

```
 4              -5              -4
```

The INT function is frequently used to round off a number X to I places, as in Example (3).

EXAMPLE (3)

PROGRAM:

```
10 PRINT INT(13.22+.5)
20 PRINT INT(7.64+.5)
30 PRINT INT(13.2246*100+.5)/100
40 PRINT "WHAT NUMBER DO YOU WANT ROUNDED OFF ";
50 INPUT X
60 PRINT "TO HOW MANY PLACES ";
70 INPUT I
80 LET N=INT(X*10↑I+.5)/10↑I
90 PRINT X;"ROUNDED OFF TO";I;"PLACES IS";N
100 END
```

OUTPUT:

```
 13
 8
 13.22
WHAT NUMBER DO YOU WANT ROUNDED OFF ?23.7246
TO HOW MANY PLACES ?3
 23.7246 ROUNDED OFF TO 3 PLACES IS 23.725
```

For example, in Steps 10 and 20 above, INT(X+.5) rounds off a number to an integer, as follows:

$$INT(13.22+.5) = INT(13.72)$$
$$= 13$$

$$INT(7.64+.5) = INT(8.14)$$
$$= 8$$

As shown in Step 30, a number can be rounded off to two places by the formula INT(X*100+.5)/100:

$$\text{INT}(13.2246*100+.5)/100 \;=\; \text{INT}(1322.46+.5)/100$$

$$=\; \text{INT}(1322.96)/100$$

$$=\; 1322/100$$

$$=\; 13.22$$

Step 80 demonstrates that, in general, INT(X*10↑I+.5)/10↑I rounds off a number X to I decimal places.

5. The Random Number Function

There are many programs which require the selection of numbers at random, especially in the fields of probability and statistics. For such problems, the BASIC language provides a function which generates random numbers: RND.

Example (4) illustrates how we can use the functions RND and INT to produce random numbers over any range. The function RND(X) produces a random number between 0 and 1. 10*RND (X) produces a random number between 0 and 10. INT(10*RND(X)) produces a random integer between 0 and 9 inclusive. INT(15*RND(X)+5) produces a random integer between 5 and 19 inclusive.

Step 100 shows that in general INT(A*RND(X)+B) produces a random number from A integers of which B is the smallest and (A+B-1) is the largest.

EXAMPLE (4)

PROGRAM:

```
10 PRINT RND(X)
20 PRINT 10*RND(Z)
30 PRINT INT(10*RND(Y))
40 PRINT INT(15*RND(W)+5)
50 PRINT
60 PRINT "WHAT IS THE SMALLEST RANDOM INTEGER YOU WANT ";
70 INPUT B
80 PRINT "FROM HOW MANY INTEGERS DO YOU WISH THE INTEGER ";
85 PRINT "TO BE CHOSEN ";
90 INPUT A
100 LET C=INT(A*RND(X)+B)
110 PRINT "A RANDOM INTEGER BETWEEN"B;"AND"A+B-1;"IS"C
120 END
```

OUTPUT:

```
.217873
6.96209
2
19
```

```
WHAT IS THE SMALLEST RANDOM INTEGER YOU WANT ?100
FROM HOW MANY INTEGERS DO YOU WISH THE INTEGER TO BE CHOSEN ?901
A RANDOM INTEGER BETWEEN 100 AND 1000 IS 517
```

On some systems the argument X of the RND function may be omitted. Step 10 of Example (4) above would then read 10 PRINT RND, Step 20 would read 20 PRINT 10∗RND, etc.

Look at Example (5). Notice that in both RUNs of the program, the same two "random" numbers were printed. These numbers are also the first two values of RND(X) in Example (4), namely, 0.217873 and 0.696209.

EXAMPLE (5)

PROGRAM:

```
10 PRINT RND(X), RND(X)
20 END
```

OUTPUT (1):

```
.217873      .696209
```

OUTPUT (2):

```
.217873      .696209
```

We can generate a different set of random numbers in each RUN by using the command RANDOM, or RANDOMIZE. In Example (6), the computer printed two different sets of random numbers because of the RANDOM instruction of Step 10.

EXAMPLE (6)

PROGRAM:

```
10 RANDOM
20 PRINT RND(X), RND(X)
30 END
```

OUTPUT (1):

.371927   .604279

OUTPUT (2):

7.37391 E-2   .390584

However, some systems always generate a different set of numbers each time the RND function is used. On these systems the command RANDOM is unnecessary.

Finally, some systems use the value of the argument to determine the function value: if $X > 0$, RND(X) always gives the same value for a given X; if $X = 0$, RND(X) supplies a standard, repeatable list of random numbers; if $X < 0$, RND(X) generates a new, unrepeatable sequence of random numbers. Check your system to determine the behavior of the RND function.

6. The Square Root Function

Step 10 of Example (7) shows three ways that the computer can calculate the square root of a number: by using the function SQR, by raising to the power (1/2), and by raising to the power .5 . The SQR function is the most efficient method of the three.

Notice in Step 20 that the computer calculates precisely what it was told to do, namely: raise to the power of 1, and then divide the result by 2.

Example (8) illustrates how the computer reacts to certain cases which are unusual or undefined.

We know that the square root of a negative number is undefined over the set of real numbers. Step 10 demonstrates that, if you ask the computer to calculate the square root of a negative number, it first gives you an ERROR message, and then calculates the square root of the absolute value of that number.

We also know that a negative number raised to a nonintegral

EXAMPLE (7)

PROGRAM:

```
10 PRINT SQR(16.4824), 16.4824↑(1/2), 16.4824↑.5
20 PRINT 16.4824↑1/2
30 END
```

OUTPUT:

```
4.05985        4.05985        4.05985
8.2412
```

power is undefined over the set of real numbers. Step 20 of Example (8) instructs the computer to calculate such a number. The machine responds with an ERROR message, and then prints the value of (ABS(-3)) ↑ 3.2.

Notice in Step 30, however, that since a negative number raised to an integral power is well-defined over the real numbers, the computer correctly evaluates the expression.

EXAMPLE (8)

PROGRAM:

```
10 PRINT SQR(-4)
20 PRINT (-3)↑3.2
30 PRINT (-3)↑3
40 END
```

OUTPUT:

```
SQUARE ROOT OF NEGATIVE NUMBER IN 10
 2
ABSOLUTE VALUE RAISED TO POWER IN 20
 33.6347
-27
```

7. The Logarithmic Function

Recall from your second-year algebra course that the logarithm of a number x to a given base b is the exponent y to which you must raise that base b in order to obtain x. In symbols, we write:

4-1 Built-in Functions   53

$\log_b x = y$ if and only if $b^y = x$,

where x is a positive real number and b is a positive real number not equal to 1.

The base b which is used on the computer is the number e, where $e \doteq 2.71828$. $\log_e x$, often written ln x, is called the natural logarithmic function, and is written LOG(X) in BASIC.

Look at Step 10 of Example (9). We know that $\ln 1 = \log_e 1 = 0$, since $e^0 = 1$. We also know that ln e = 1, since $e^1 = e$. Thus, 2.71828 must be a rather good approximation for e, since the computer's value of LOG(2.71828) is .999999, which is certainly very close to 1.

EXAMPLE (9)

PROGRAM:

```
10 PRINT LOG(1), LOG(2.71828)
20 END
```

OUTPUT:

```
 0              .999999
```

In Step 10 of Example (10), we received an appropriate ERROR message, since LOG(X) is well-defined only when X is a positive real number. However, the computer went on to evaluate LOG(ABS(-1)).

EXAMPLE (10)

PROGRAM:

```
10 PRINT LOG(-1)
20 PRINT LOG(0)
30 END
```

OUTPUT:

```
LOG OF NEGATIVE NUMBER IN 10
 0
LOG OF ZERO IN 20
-1.70141 E+38
```

In Step 20 of Example (10), the computer printed another appropriate ERROR message, since, as x approaches 0 from the right, ln x approaches negative infinity. The computer tried to accommodate us by printing its value for negative infinity: -1.70141×10^{38}.

8. The Exponential Function

An exponential function is of the form b^x, where b is any positive real number and x is any real number. If we let the base b be the number e, then we have a rather special exponential function e^x. It is this exponential function which is incorporated into the BASIC language. It is written EXP(X) on the computer.

The OUTPUT of Step 10 of Example (11) shows that indeed the base b of EXP(X) is e, since $e^1 = e$.

Step 20 illustrates the fact that EXP(X) and LOG(X) are *inverse* functions. Recall from your work in mathematics that

$$e^{\ln x} = x \quad \text{and} \quad \ln e^x = x.$$

In BASIC, we write:

$$\text{EXP(LOG(X))} = X \quad \text{and} \quad \text{LOG(EXP(X))} = X.$$

You see one special case of the above equations in Step 20.

EXAMPLE (11)

PROGRAM:

```
10 PRINT EXP(1)
20 PRINT EXP(LOG(17)), LOG(EXP(17))
30 END
```

OUTPUT:

```
2.71828
17.            17.
```

---

9. The Trigonometric Functions

The trigonometric functions included in BASIC are sine (written SIN on the computer), cosine (COS), tangent (TAN), and arctangent (ATN). Some systems also include the cotangent function (COT).

4-1 Built-in Functions 55

The argument X of SIN(X), COS(X), TAN(X), and COT(X) is measured in a unit of circular measure called a *radian*. One radian is approximately 57°17′. Hence, in Step 20 of Example (12), the computer printed exactly as it was directed (i.e., the sine of 60 radians instead of the sine of 60 degrees, as intended). If you want to calculate the sine of 60 degrees, then you must first convert from degree measure to radian measure, using the formula:

$$m^R(\alpha) = \frac{\pi}{180} \cdot m^o(\alpha) \quad ,$$

where $m^R(\alpha)$ means the measure of angle α in radians, and $m^o(\alpha)$ means the measure of angle α in degrees. Using the above formula for $\alpha = 60$, we have:

$$m^R(60) = \frac{\pi}{180} \cdot 60$$

$$= \frac{\pi}{3} \ .$$

Thus, using the value assigned to P in Step 10, the computer calculates and prints the sine of $\pi/3$ radians or 60 degrees in the second part of Step 20.

EXAMPLE (12)

PROGRAM:

```
10 LET P=3.14159
20 PRINT SIN(60), SIN(P/3)
30 PRINT TAN(P/2)
40 PRINT COT(P/3), COS(P/3)/SIN(P/3), 1/TAN(P/3)
50 PRINT ATN(1), P/4
60 PRINT ATN(SQR(3)), P/3
70 END
```

OUTPUT:

```
-.304810        .866025
 736601.
 .577351        .577351        .577351
 .785398        .785397
1.04720        1.04720
```

Step 30 of Example (12) above illustrates an attempt to calculate TAN($\pi/2$). From your work in mathematics, you know that, as X approaches $\pi/2$ from the left, TAN(X) becomes arbitrarily large. At precisely $\pi/2$, the tangent function is undefined. Since our value for π in Step 10 is only accurate to six significant digits, the computer is actually calculating the tangent of a number very close to $\pi/2$; it thus prints the very large number 736601. for the value of TAN($\pi/2$).

Step 40 calculates COT($\pi/3$) in three ways: by using the built-in COT function, and from the formulas

$$\cot x = \frac{\cos x}{\sin x}, \qquad \text{where } \sin x \neq 0, \text{ and}$$

$$\cot x = \frac{1}{\tan x}, \qquad \text{where } \tan x \neq 0.$$

Steps 50 and 60 illustrate the use of the arctangent function. The arctangent function ATN is the inverse of the tangent function TAN. The argument X of ATN(X) can be any real number, and the image element ATN(X) may be thought of as a radian between $-\pi/2$ and $\pi/2$. You know that Arctan $1 = \pi/4$ and that Arctan $\sqrt{3} = \pi/3$. Six-digit approximations of the numbers $\pi/4$ and $\pi/3$ are given in the OUTPUT of Steps 50 and 60, along with the values ATN(1) and ATN(SQR(3)). Notice that these numbers are almost identical.

The reason the secant, the cosecant, and, on some systems, the cotangent functions are not included in BASIC is that they can be computed from the built-in trigonometric functions. For example, you have already seen in Step 40 of Example (12) two ways in which COT(X) can be calculated using the SIN, COS, and TAN functions. You can just as easily find the values of secant x and cosecant x from the identities:

$$\sec x = \frac{1}{\cos x}, \qquad \text{where } \cos x \neq 0, \text{ and}$$

$$\csc x = \frac{1}{\sin x}, \qquad \text{where } \sin x \neq 0.$$

The arcsine and arccosine functions can be calculated from the arctangent function, as you will learn in Section 7-2.

EXERCISES 4-1

Punch a tape of the following problem, LOG ON to the computer, feed your tape through the reader, RUN the program, then LOG OUT.

4-1 Built-in Functions   57

```
1. 10   PRINT "COT:" COT(3.14159/6)
   20   END

2. 10   RANDOM
   20   PRINT "ABS:"ABS(-35); ABS(3.14); ABS(-2.48)
   30   PRINT "SQR:"SQR(2); 2 ↑ (1/2); 2 ↑ 1/2; SQR(-2)
   40   PRINT "INT:" INT(37.4672); INT(5); INT(-2.4168); INT(-5)
   50   PRINT "SGN:" SGN(5.36); SGN(-3.95); SGN(0)
   60   PRINT "LOG:" LOG(2.71828183); LOG(1); LOG(-3.14)
   70   PRINT "EXP:" EXP (1); EXP(0); EXP(-2.56); EXP(5.63)
   80   PRINT "LOG AND EXP:" LOG(EXP(3)); EXP(LOG(3)); EXP(LOG(5));
   90   PRINT LOG(EXP(-5))
   100  LET P = 3.14159
   110  PRINT "SIN:" SIN(30); SIN(30*P/180); SIN(P/6)
   120  PRINT "COS:" COS(P/2); COS(P/3); COS(P/6)
   130  PRINT "TAN:" TAN(0); TAN(P/4); SIN(P/4)/COS(P/4)
   140  PRINT "ATN:" ATN(1); ATN(SQR(3))
   150  PRINT "RND:" RND(X); 10 *RND(X)+10
   160  PRINT "RND AND INT:" INT(20*RND(X)+5); INT(-10*RND(X)-10)
   170  PRINT "MIXED:" SIN(COS(LOG(P))); SQR(SIN(INT(1.7)*LOG(3)))
   180  END
```

4-2 Numbers in BASIC

In this section we shall discuss the kinds of numbers which are accepted in BASIC, the possible magnitude of these numbers, and the form in which the computer prints them.

1. Numbers Accepted in BASIC

The BASIC language accepts numbers written in integral (i.e., 137, -2, -123456789), decimal (i.e., 146.28362, -.177), or exponential (i.e., 1.36E-5, -236E8) form. Numbers may be either positive, negative, or zero, but must not contain commas. For example, BASIC does not accept 12,476; the number should be written 12476.

On most systems numbers may not contain more than nine digits. Thus your system might not accept .0123456789, because it contains ten digits. This presents no serious problem, however, since you can write such a number in exponential form: 1.23456789E -2, or 12345.6789E-6, or 123456789E-10.

2. Range and Accuracy

On most systems BASIC is capable of handling numbers ranging from $\pm 1.49637 \times 10^{-39}$ to $\pm 1.70141 \times 10^{38}$. Also, on most systems BASIC performs all its calculations with nine significant digits.

3. OUTPUT Numbers

Most BASIC systems follow these rules for printing numbers.

a. *Integers*. Integers up to eight digits long are printed in integral form without a decimal point, as you can see in the OUTPUT of Step 10 of Example (1). Integers containing more than eight digits are rounded off to six digits and printed in exponential form, as you can see in the OUTPUT of Step 20.

EXAMPLE (1)

PROGRAM:

```
10 PRINT 12345678
20 PRINT 123456789
30 PRINT 463.7200, 0.013294
40 PRINT 0.0132946
50 END
```

OUTPUT:

```
12345678
1.23457 E+8
463.72           .013294
1.32946 E-2
```

b. *Decimals*. Up to six significant digits are printed for any decimal. Any decimal which can be printed in less than seven digits is printed as a decimal, as Step 30 of Example (1) above illustrates. Any decimal which cannot be printed in six digits is typed in exponential format, as in the OUTPUT of Step 40.

Also notice in the OUTPUT of Step 30 that trailing zeros after the decimal point are not printed.

EXERCISES 4-3

Type problems 1 and 2 OFF LINE, LOG ON to the computer, feed the tape through the reader, and then LOG OUT.

1. ```
   10 PRINT 123456789
   20 PRINT 1E40
   30 PRINT 1E80
   40 END
   ```
   ```

2. 10 PRINT 16E5, 2.176E30, 2E+30, 1E-5, 1E-1
   20 PRINT 99999991+8, 99999991+9, 1E10
   30 PRINT 0.2456004, 0.2456007, 0.0024561
   40 PRINT EXP(88.02)
   50 PRINT EXP(88.03)
   60 END

4-3 Summary

In this chapter you have learned about the functions incorporated into the BASIC language. These functions have been provided in order to make more effective use of the computer for solving problems. Although most of the examples used thus far to illustrate these built-in functions were elementary in nature, later on in the book you will see more sophisticated programs which use these functions to great advantage. Also, in Section 6-4 you will learn how to define and use your own functions in a program.

You have also learned about the kinds of numbers the computer can accept, work with, and print in the BASIC language.

An appreciation of the kinds of numbers used in BASIC, along with a good understanding of the built-in functions, will greatly increase your versatility as a programmer.

CHAPTER 5

Now that you have some experience and familiarity with the Teletype console, can punch a tape, LOG IN and OUT, and can use some of the built-in functions to calculate rather complicated expressions, you are ready to take advantage of the real power of the computer by writing more sophisticated and more useful programs.

Many problems cannot be solved by programs which consist only of a linear sequence of statements. Some programs require a branching technique which alters the linear order of execution of the steps in a program. In this chapter you will learn how to change the order of steps by means of the GO TO command. You will also learn how to use the IF-THEN command — that command which directs the computer to another step in a program, dependent upon certain stated conditions.

A problem often requires a program which repeats the execution of a few steps a number of times. This repetitive technique is made quite simple in BASIC by means of the FOR-NEXT commands.

In this chapter you will also make flow charts, and you will begin to appreciate the ease with which you can write a program with the help of a flow chart. You will also become familiar with the alphanumeric data and string variable capabilities of the BASIC language.

These ideas and commands are presented within the context of particular programs. It is hoped that this approach will make the ideas meaningful to you.

5-1 Making a Flow Chart

Suppose you are asked to write a program which will find the area A of a circle and the volume V of a sphere for a given radius R.

You can organize your thoughts about the program by first making what is called a *flow chart*. A flow chart is simply a schematic device which outlines the procedure by which the computer solves a problem.

The following table of symbols, with explanation, will be used when flow charts are constructed in this book.

Table (1). FLOW CHART SYMBOLS

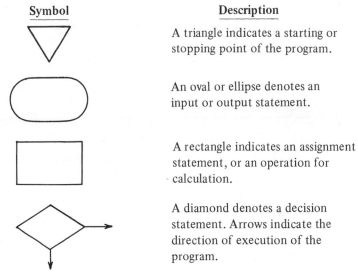

Symbol	Description
▽	A triangle indicates a starting or stopping point of the program.
⬭	An oval or ellipse denotes an input or output statement.
▭	A rectangle indicates an assignment statement, or an operation for calculation.
◇	A diamond denotes a decision statement. Arrows indicate the direction of execution of the program.

For example, the flow chart for the area and volume problem given above is as follows.

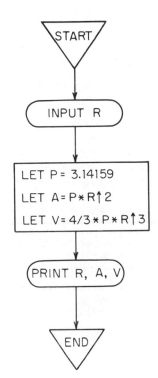

Figure (1)

The arrows in Figure (1) indicate the direction the flow chart is read. The order in which the geometrical shapes appear in the flow chart is dictated by the order in which the computer ordinarily executes the program. For example, the computer must have a value for R before it can determine A and V. Thus the INPUT R statement must precede the LET statements in the flow chart.

EXERCISE 5-1

Make a flow chart of the following problems:

1. Input two numbers R and H, and then type the volume and total surface area of a right circular cylinder with radius R units and height H units.
2. Input a number X, and then type the number X, the square of X, and the square root of X.
3. Input three numbers A, B, and C, and then type the average M of those three numbers.

5-2 Program (1): The Area of a Circle and the Volume of a Sphere (1)

1. Program (1)

Using the flow chart given in Section 5-1 above, we can write the following program.

PROGRAM (1):

```
10 PRINT "   THIS PROGRAM COMPUTES THE AREA OF A CIRCLE AND THE"
20 PRINT "VOLUME OF A SPHERE FOR ANY RADIUS."
30 PRINT
40 PRINT "WHAT IS THE RADIUS ";
50 INPUT R
60 LET P=3.14159
70 LET A=P*R↑2
80 LET V=4/3*P*R↑3
90 PRINT
100 PRINT " RADIUS"," AREA"," VOLUME"
110 PRINT R,A,V
120 END
```

5-2 Program (1): The Area of a Circle and the Volume of a Sphere (1)   63

OUTPUT:

```
THIS PROGRAM COMPUTES THE AREA OF A CIRCLE AND THE
VOLUME OF A SPHERE FOR ANY RADIUS.

WHAT IS THE RADIUS ?5.382

   RADIUS          AREA             VOLUME
   5.382           90.9991          653.009
```

2. Comments about Program (1)

a. As we have stated previously, the computer performs each step in ascending numerical order. Thus, if we typed Step 50 after Step 80 in Program (1), the computer would have rearranged the step numbers in numerical order before executing the program. If you forgot to include a step between Step 50 and 60, say, you could type after Step 120 a step numbered 52. The computer would put Step 52 in proper numerical order before executing the program. Thus you can see that it is advisable to leave a sufficiently large increment between your step numbers so that you can insert additional steps if necessary.

b. Steps 10 and 20 give a description of the program. Step 40 asks for the radius. The semicolon at the end of Step 40 suppresses a carriage return and line feed and causes the ? from the **INPUT** statement of Step 50 to be printed immediately after the word RADIUS.

c. Step 60 sets P equal to 3.14159, an approximate value for π. Steps 70 and 80 set the values of A (area) and V (volume) according to formulas with which you are familiar.

d. Step 100 prints the heading for each column, and Step 110 types the values of R, A, and V. Recall that the first space of each column is reserved for the sign of the number. If the number is positive, the space is blank; if the number is negative, a negative sign is printed in the space. In the OUTPUT of Step 110, the first space of each column is left blank, since the values of R, A, and V are positive numbers. For this reason, we have included one space at the beginning of each string in Step 100. Notice in the OUTPUT that in each column the first digit of each number is directly under the first letter of each heading.

e. Step 120 is the END command. Recall that this statement must have the largest step number in the program. Note that some systems do not require the END statement in a program.

f. Each of Steps 30 and 90 generates a line feed and is included in the program to make the OUTPUT look a bit nicer.

3. The Execution of a Program

Recall from Chapters 3 and 4 that after you type a program OFF LINE, LOG onto the computer, and feed your tape through the tape reader, you are ready to execute your program.

When the tape of Program (1) is fed into the computer, the machine will stop after Step 120 and will await your next instruction. You then type RUN ®. (Recall that the symbol ® is used throughout this book to indicate that you strike the RETURN key.) Most systems are designed in such a way that the computer then types a heading consisting of the program name, the time, and the date.

If there are any errors in your program, the computer will type the step number and an ERROR message indicating the kind of error in that step. If there are no errors, the computer will execute the program for one value of the radius, and then type READY or the time used to execute the program.

If you wish to run the program again, simply type RUN ®. The computer will execute the entire program a second time. You can continue executing the program as many times as you want (each time typing in a different value for the radius R) by typing RUN ® after the completion of each run.

Program (2), which will be described in the next section, includes a few steps in Program (1) and is designed to perform the repetitive function much more elegantly than is done in Program (1).

4. The Procedure after the Execution of a Program

After you have executed your program a number of times and are assured that the program works as you intended, then you can do one of the following.

(a) LOG out. If you LOG out of the computer at this stage, however, you automatically erase your entire program from the computer's memory. Of course, if the copy of the program which you have on tape is your finished product, you can always feed the tape into the computer a second time.

(b) Make a second, finished copy of your program on tape. Assume that your first tape is not precisely as you want it. You have made a few minor but necessary changes while ON LINE and now want a complete and accurate second tape of your program. You can make a second tape by following this procedure.

5-2 Program (1): The Area of a Circle and the Volume of a Sphere (1)

(i) Type the command LISTNH. This command tells the computer to list the program with no heading. If you want a heading to appear at the beginning of the OUTPUT of your program, then type simply LIST. (See Chapter 11 for a complete description of the LIST commands.)

(ii) Turn the tape punch mechanism ON.

(iii) Strike the space bar or the RUBOUT key about 10 times to put some leader on the tape.

(iv) Strike the RETURN key.

As the computer types each step of your program in numerical order, it is at the same time punching a tape of the program. However, this second tape has the LINE FEED directions punched on it each time the computer finishes a line and goes on to the next.

On some systems the word READY is typed after the program has been completely listed. The appropriate holes for the word READY will also be punched on the tape. If you wish, you can delete the word READY from your tape by erasing the associated punch holes. Just follow this procedure:

(i) turn the tape punch mechanism OFF

(ii) LOG OUT

(iii) press the LOCAL button

(iv) turn the tape punch mechanism ON

(v) press the BACK SPACE button 7 times (1 time for LINE FEED, 1 for RETURN, and 5 for READY)

(vi) strike the RUBOUT key at least 7 times. The series of punched holes which correspond to the seven characters are now deleted from the tape.

(c) File your program onto disc storage by typing SAVE ®. (A *disc* is a piece of peripheral equipment on some systems which increases the storage capacity of the system.) The command SAVE transfers a copy of the program from the working storage area of the computer to the permanent storage area (the disc). Of course, if your computer does not include disc storage capabilities, then you do not have this option of filing your programs.

When the computer types READY, you know that it has filed your program onto the disc. The name of the file is the name you typed when the computer asked you NEW PROGRAM NAME at the time you first LOGged ON. On most systems the name of the program may have a maximum of six characters.

After the computer types READY, you can then LOG OUT.

Suppose after a few days you want to execute your program again. Or perhaps you stored an unfinished program on the disc and now want to continue working on it. The procedure for retrieving your program from disc storage is as follows.

(i) After LOGging ON, when the computer asks you NEW OR OLD, type OLD.

(ii) The computer then types OLD PROGRAM NAME. You type the name of the file under which you previously filed the program.

The computer will copy your program from the disc to the working storage area. When the program is completely copied, the computer will type READY. If the computer types the ERROR message PROGRAM NOT SAVED, it is telling you that you are trying to load a file which does not exist on the disc.

The format of the "loading" procedure varies with each system. A typical print-out of the loading procedure is:

NEW OR OLD--<u>OLD</u>   ®
OLD PROGRAM NAME--<u>SPHER1</u>   ®

<u>READY</u>

Note that what is underlined in the above print-out was typed by the computer.

We suggest that you become familiar with the system and the EDIT commands of your particular system, as given in Chapter 11 and in the manual designed for your system. Knowing the system and EDIT commands will greatly facilitate your programming expertise.

EXERCISES 5-2

Perform each of the following steps for each of the following programs, one program at a time.

a) Construct a Flow Chart for your program.
b) Write the program.
c) Punch a tape OFF LINE.
d) LOG ON, and feed your tape into the computer.
e) If the program works, execute a few and critical values.
f) Make a LISTNH copy of your finished program, at the same time punching a tape of the program.
g) LOG OUT.

1-3. Problems 1-3, Section 5-1.

4. Input a number N of inches, convert that number of inches into the corresponding number of centimeters, and type the result.

5-2 Program (1): The Area of a Circle and the Volume of a Sphere (1)

5. Input the lengths of two sides of a right triangle and then type out two possible lengths of the third side.

6. Input two numbers, A and D, and then type the remainder of A divided by D.

5-3 The IF-THEN Command; Program (2): The Area of a Circle and the Volume of a Sphere (2)

1. The IF-THEN Command

The IF-THEN command permits us to create a conditional branching of a program. The general format of the IF-THEN statement is:

IF (expression) (relation) (expression) THEN (step number).

Both expressions are evaluated and compared by the relation in the statement. If the condition is true, the computer jumps to the step given after THEN. If the condition is false, the computer continues to the next step of the program.

In IF-THEN statements, six relation symbols are used to compare values.

Table 2. RELATIONS IN BASIC

Symbol	Relation
=	equal to
<	less than
>	greater than
<=	less than or equal to
>=	greater than or equal to
<>	not equal to

The number following the word THEN may on most systems be the line number of any step in the program. However, on some systems control may not be transferred to the nonexecutable steps of DATA, DEF (See Section 6-4), DIM (see Section 5-8) and REM (see Section 6-1). If the number following the word THEN is not a step number in your program, the computer will type an ERROR message.

Consider the following example.

EXAMPLE (1)

PROGRAM:

```
10 PRINT "TYPE A NUMBER ";
20 INPUT N
30 PRINT N;"IS ";
40 IF N/2=INT(N/2) THEN 60
50 PRINT "NOT ";
60 PRINT "EVEN."
70 END
```

OUTPUT(1):

```
TYPE A NUMBER ?16
 16 IS EVEN.
```

OUTPUT (2):

```
TYPE A NUMBER ?13
 13 IS NOT EVEN.
```

In Step 40 of Example (1), the relation of equality was used in the IF-THEN statement to determine whether N is even or odd. If N is even, it is divisible by 2; hence, $N/2 = INT(N/2)$. In OUTPUT (1), since the conditional part of Step 40 was satisfied, the computer jumped to Step 60. In OUTPUT (2), since the conditional was not satisfied (i.e., $N/2 <> INT (N/2)$), the computer went to the next step.

2. The Compilation of a Program

The IF-THEN statement provides a more elegant way than heretofore of having the computer repeat the execution of a program without our having to type RUN. Each time we type RUN, the computer must *compile* the program and then begin execution from the first step. Compilation of the program is the process by which the computer translates the program from the BASIC language into machine language — that language which the computer understands. Compilation of the program after the first RUN consumes unnecessary computer time. Also, having the computer type a complete description of a familiar program wastes both our time and the computer's. The IF-THEN statement provides a method of avoiding this unnecessary duplication.

3. The Flow Chart of Program (2)

A flow chart of this more elegant approach to Program (1) is as follows.

5-3 The IF-THEN Command; Program (2)   69

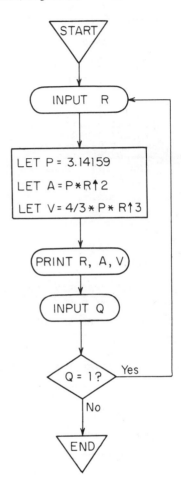

Figure (2)

The important distinction between the above flow chart and the one in Figure (1) is the diamond-shaped *decision box*. We want the computer to ask us if it should repeat the program. If we type the number 1 for Q, then the program will repeat; if we type any other number, then the computer will stop.

4. Program (2): The Area of a Circle and the Volume of a Sphere (2)

Program (2) is as follows.

PROGRAM (2):

```
10 PRINT "    THIS PROGRAM COMPUTES THE AREA OF A CIRCLE AND THE"
20 PRINT "VOLUME OF A SPHERE FOR ANY RADIUS."
30 PRINT
40 PRINT "WHAT IS THE RADIUS ";
50 INPUT R
60 LET P=3.14159
70 LET A=P*R↑2
80 LET V=4/3*P*R↑3
90 PRINT
100 PRINT " RADIUS", " AREA", " VOLUME"
110 PRINT R,A,V
120 PRINT "AGAIN (TYPE 1 FOR YES, 0 FOR NO) ";
130 INPUT Q
140 IF Q=1 THEN 30
150 END
```

OUTPUT:

```
   THIS PROGRAM COMPUTES THE AREA OF A CIRCLE AND THE
VOLUME OF A SPHERE FOR ANY RADIUS.

WHAT IS THE RADIUS ?453.92

 RADIUS          AREA            VOLUME
 453.92          647303          3.91766 E+08
AGAIN (TYPE 1 FOR YES, 0 FOR NO) ?1

WHAT IS THE RADIUS ?.1425

 RADIUS          AREA            VOLUME
 .1425           6.37939 E-02    1.21208 E-02
AGAIN (TYPE 1 FOR YES, 0 FOR NO) ?0
```

5. Comments about Program (2)

a. The significant change in Program (2) from Program (1) is in Step 140, in which the conditional IF-THEN is used. During the execution of the program and after the computer calculates and types A and V for a given value of R, it then types AGAIN (TYPE 1 FOR YES, 0 FOR NO), from Step 120. The computer then types a ?, from Step 130. If you type 1 for Q, then in Step 140 the computer "reads" the IF statement, realizes that the conditional is satisfied, i.e., Q = 1, and executes the THEN part of the statement by jumping back to Step 30.

If you type 0 for Q in Step 130, then, in Step 140, the computer reads the IF clause, realizes that the conditional is not satisfied, i.e.,

Q <> 1, ignores the THEN part of Step 140, and proceeds to the next step (150). Note that any number other than 1 which is typed after ? will cause the computer to continue to Step 150.

After the computer reaches Step 150, it types READY, thus awaiting your next instruction. Your next command might be a LOG OUT instruction, a LISTNH instruction, a SAVE instruction, or some other system command.

    b. In Step 140 of Program (2), we could have written IF Q > 0 THEN 30. Thus, if you typed any positive number for Q, the computer would have jumped to Step 30. If you had typed any nonpositive number for Q, the program would have ended.

    c. On some systems, the command IF-GO TO can be used in place of the IF-THEN command. The two statements are equivalent.

    d. It would be nice if we could input *words* such as YES or NO instead of *numbers* in Step 130. Indeed, the BASIC language has a feature which enables us to input, print, store, and compare words. In the next section we shall discuss this feature of BASIC.

EXERCISES 5-3

1-6. Change each of the six programs you wrote in Exercises 5-2 to provide for the repetition of each program. Use the IF-THEN statement.

5-4 Alphanumeric Data and String Variables

1. Strings and String Variables

A *string* is a sequence of digits, letters, or other noncontrol characters of a system. On some systems a string may contain a maximum of sixty characters. On others strings are limited to fifteen characters. However, on all systems, a string may not contain quotation marks.

    A *string variable* is any variable followed by a dollar sign ($).

    Look at Example (1). We have assigned a string to each of the given string variables by means of the LET command. Step 10 assigns the string BASIC to the string variable A$, Step 20 assigns the string ALPHANUMERIC to the string variable B1$, etc. Notice that the string must be enclosed in quotation marks.

    If a subscripted variable is used as a string variable, then the $

must precede the subscript, as you can see in Steps 30 and 40 of Example (1). Step 40 assigns the string VARIABLES to two variables: D$(5) and E$(1). Note that we may leave out the word LET, as Steps 30 and 40 illustrate.

Notice that string variables are printed exactly like strings enclosed in quotation marks: a comma causes the strings to be printed in the 5-column arrangement and a semicolon prints the strings with no spaces between them. Also, any combination of numeric and string variables and strings enclosed in quotation marks may appear in a PRINT statement, separated by commas or semicolons. (No punctuation between a string in quotation marks and a string variable prints the strings with no spaces between them.)

On some systems a double letter notation is used instead of the $ notation for string variables. For example, AA would be used instead of A$, and DD(5) instead of D$(5).

EXAMPLE (1)

PROGRAM:

```
10 LET A$="BASIC"
20 LET B1$="ALPHANUMERIC"
30 C$(2)="STRING 123456789"
40 D$(5)=E$(1)="VARIABLES"
50 PRINT A$;B1$,C$(2),D$(5),E$(1)
60 END
```

OUTPUT:

```
BASICALPHANUMERIC        STRING 123456789            VARIABLES
VARIABLES
```

2. Renaming a String Variable

Look at Example (2). In Step 10 we have assigned the string JANUARY 1 to the string variable A$, and then in Step 20 we assigned the contents of A$ to the string variable B$(5). This technique of renaming a string variable is quite useful in some programs.

EXAMPLE (2)

PROGRAM:

```
10 LET A$="JANUARY 1"
20 LET B$(5)=A$
30 PRINT B$(5)
40 END
```

OUTPUT:

JANUARY 1

3. READ-DATA and INPUT Statements with String Variables

Strings may also be assigned to string variables by means of READ-DATA and INPUT statements. In DATA statements, strings which do not begin with a letter or strings which contain commas or leading blank spaces must be enclosed in quotation marks. Trailing blank spaces in unquoted strings are considered significant characters. Thus in Step 40 of Example (3) we have placed DEC. 7, 1941 in quotes because it contains a comma; 4+H is in quotation marks because it does not begin with a letter; and the string   XYZ is enclosed in quotes because it contains two leading blank spaces. Notice in the OUTPUT that the computer considered the three spaces following HYDROGEN part of the string A$, but that it ignored the two blank spaces before ABC.

In INPUT statements, strings containing commas or leading blank spaces must be enclosed in quotation marks. All trailing spaces are considered significant characters. Notice in Example (3) that we placed the string 10,000 in quotation marks because it contains a comma. Also notice that the four blank spaces following OXYGEN are counted as part of the string, and that it was not necessary to enclose 12 MONTHS in quotation marks, regardless of the fact that the first character is a number.

When in doubt about using quotation marks around strings, by all means use them. The use of quotation marks around any kind of string will not cause an ERROR message to be printed.

EXAMPLE (3)

PROGRAM:

```
10 READ A$,B1$,K$,F$(2),G$
20 INPUT Q$(3),S$,P$
30 PRINT A$;B1$,K$;F$(2);G$,Q$(3);S$,P$
40 DATA HYDROGEN   ,"DEC. 7, 1941","4+H","  XYZ",  ABC
50 END
```

OUTPUT:

```
?OXYGEN    ,"10,000",12 MONTHS
HYDROGEN   DEC. 7, 1941        4+H   XYZABC     OXYGEN    10,000
12 MONTHS
```

4. Mixing Numeric and String Data

Numeric and string data may be mixed in DATA statements, as in Example (4). When the DATA are read, the computer assigns the numbers to the numerical variables and the strings to the string variables in the order in which they appear in both the READ and DATA statements. Thus in Step 10 the computer sets A1 = 21, B(5) = 9, F$(1) = "ADDITION ", C6 = 10, G2$ = "(+) OF" and Q$(4) = " 3 NUMBERS:"

Notice that the space between the strings ADDITION and (+) OF is part of the string F$ (1), since there is a blank following the word ADDITION in the DATA statement (Step 20). Also notice that the space between (+) OF and 3 NUMBERS: is part of the string Q$ (4), since the blank preceding the 3 is enclosed in quotation marks.

Step 40 illustrates that we may also mix numeric and string variables in INPUT statements. In Example (4) we entered 52 for Y and 1 YEAR for W$. Notice the format of the PRINT statement in Step 50.

EXAMPLE (4)

PROGRAM:

```
10 READ A1,B(5),F$(1),C6,G2$,Q$(4)
20 DATA 21,ADDITION ,9,10,"(+) OF","  3 NUMBERS:"
30 PRINT F$(1);G2$;Q$(4);A1+B(5)+C6
40 INPUT Y,W$
50 PRINT Y;"WEEKS = "W$
60 END
```

OUTPUT:

```
ADDITION (+) OF 3 NUMBERS: 40
?52,1 YEAR
  52 WEEKS = 1 YEAR
```

5. Command RESTORE with Alphanumeric Data

Study very carefully the OUTPUT of Example (5). The RESTORE command in Step 20 restores *both* the numeric and string data. Thus in the OUTPUT you see the second 1 for the variable D and the second AB for S$. The command RESTORE* in Step 40 restores only the numeric data. That is why you see the third 1 for the variable F in the OUTPUT, and the E for R$. The command RESTORE$ in Step 60 restores only the string data. Thus O$ is assigned the string AB, while H is set equal to 3.

Notice that the semicolons in Step 80 caused the strings to be printed in the OUTPUT with no spaces between them.

EXAMPLE (5)

PROGRAM:

```
10 READ A,B,C,X$,Y$,Z$
20 RESTORE
30 READ D,E,S$,T$
40 RESTORE*
50 READ F,G,R$
60 RESTORE$
70 READ H,O$,P$,Q$
80 PRINT A;D;F;H;X$;S$;R$;O$
90 DATA 1,2,AB,3,CD,E
100 END
```

OUTPUT:

```
 1  1  1  3 ABABEAB
```

6. The IF-THEN Command with String Variables

String variables may also appear in IF-THEN statements, as in Example (6). In Step 30 of Example (6), the program stops if we answer EASY.

If we answer HARD (or any string other than EASY), the computer continues to Step 40 and prints OH, YOU'RE FOOLING!.

Note that the string EASY must be enclosed in quotation marks in Step 30.

EXAMPLE (6)

PROGRAM:

```
10 PRINT "IS THIS EASY OR HARD ";
20 INPUT E$
30 IF E$="EASY" THEN 50
40 PRINT "OH, YOU'RE FOOLING!"
50 END
```

OUTPUT (1):

```
IS THIS EASY OR HARD ?HARD
OH, YOU'RE FOOLING!
```

OUTPUT (2):

```
IS THIS EASY OR HARD ?EASY
```

The six relation symbols $=, <, >, <=, >=, <>$ can be used with strings and string variables to make alphabetical comparisons. Thus if ARCSIN is the string assigned to F$ and ARCCOS is the string assigned to G$, then G$ < F$, since ARCCOS precedes ARCSIN alphabetically.

Hence we can change Step 30 of Example (6) to the statement IF E$ < ="E" THEN 50, as in Step 30 of Example (7). Any word beginning with A, B, C, D or E will cause the computer to jump to Step 50. Any word beginning with a letter from F to Z will cause the computer to continue to Step 40, and the string OH, YOU'RE FOOLING! will thus be printed, as you can see in OUTPUT (1).

7. The Comparison of String Variables

We may use either a string or a string variable on either side of the relation symbol in alphanumeric IF-THEN statements. In Example (8), the computer compares two string variables A$ and B$ and determines whether or not they are the same.

In making string comparisons, the computer ignores trailing spaces. For example, the computer considers "HORSE" to be the same as "HORSE   ".

EXAMPLE (7)

PROGRAM:

```
10 PRINT "IS THIS EASY OR HARD ";
20 INPUT E$
30 IF E$<="E" THEN 50
40 PRINT "OH, YOU'RE FOOLING!"
50 END
```

OUTPUT (1):

```
IS THIS EASY OR HARD ?Q
OH, YOU'RE FOOLING!
```

OUTPUT (2):

```
IS THIS EASY OR HARD ?E
```

EXAMPLE (8)

PROGRAM:

```
10 PRINT "TWO WORDS ";
20 INPUT A$,B$
30 IF A$<>B$ THEN 50
40 PRINT "THE WORDS ARE THE SAME."
50 END
```

OUTPUT (1):

```
TWO WORDS ?HORSE,HOUSE
```

OUTPUT (2):

```
TWO WORDS ?HORSE,HORSES
```

OUTPUT (3):

```
TWO WORDS ?HORSE,HORSE
THE WORDS ARE THE SAME.
```

8. Program (3): The Area of a Circle and the Volume of a Sphere (3)

In the following Program (3) we have used string variables in Steps 130 and 140 so that we can type the words YES or NO instead of 1 or 0 in response to the question DO YOU WISH TO RUN AGAIN ?.

PROGRAM (3):

```
10 PRINT "   THIS PROGRAM COMPUTES THE AREA OF A CIRCLE AND THE"
20 PRINT "VOLUME OF A SPHERE FOR ANY RADIUS."
30 PRINT
40 PRINT "WHAT IS THE RADIUS ";
50 INPUT R
60 LET P=3.14159
70 LET A=P*R↑2
80 LET V=4/3*P*R↑3
90 PRINT
100 PRINT " RADIUS", " AREA", " VOLUME"
110 PRINT R,A,V
120 PRINT "DO YOU WISH TO RUN AGAIN ";
130 INPUT Q$
140 IF Q$="YES" THEN 30
150 END
```

OUTPUT:

```
    THIS PROGRAM COMPUTES THE AREA OF A CIRCLE AND THE
VOLUME OF A SPHERE FOR ANY RADIUS.

WHAT IS THE RADIUS ?1

   RADIUS         AREA           VOLUME
   1              3.14159        4.18879
DO YOU WISH TO RUN AGAIN ?YES

WHAT IS THE RADIUS ?5743.89

   RADIUS         AREA           VOLUME
   5743.89        1.03648 E+8    7.93792 E+11
DO YOU WISH TO RUN AGAIN ?NO
```

EXERCISES 5-4

1-6. Modify each of the programs you wrote in Exercises 5-3 so that typing YES or NO in response to DO YOU WISH TO RUN AGAIN ? causes the program to repeat or not repeat.

5-5 The GO TO Command; Program (4): The Real Roots of $AX^2+BX+C = 0$

1. The GO TO Command

The GO TO Command instructs the computer to branch unconditionally to another step in a program. The general format of the statement is:

GO TO (Step number).

In executing a GO TO statement, the computer jumps to the step whose number appears in the statement. The computer continues to execute the program from that step number to which it was sent in the GO TO command.

The step number appearing in a GO TO statement may on most systems be the number of any statement in the program. On some systems control may not be transferred to the nonexecutable statements of DATA, DEF (see Section 6-4), DIM (see Section 5-8) and REM (see Section 6-1). Also, if the number following the GO TO command is not a line number in the program, the computer will respond with an appropriate ERROR message.

Before we introduce the flow chart and program which illustrate the use of the GO TO statement, we offer some mathematical background to help you to appreciate the program.

2. The Real Roots of $AX^2+BX+C = 0$

Near the end of your first-year algebra course, you learned that the real roots of the general quadratic equation $AX^2+BX+C = 0$, where $A \neq 0$, are given by:

$$X = \frac{-B \pm \sqrt{B^2-4AC}}{2A}$$

Let us use this fact to write a program which will ask for the values of A, B, and C, and then compute and type the two real roots, if they exist. We will utilize in this program most of the ideas you have learned thus far. First we shall construct the flow chart.

3. The Flow Chart of Program (4)

Study very carefully the following flow chart of the above problem.

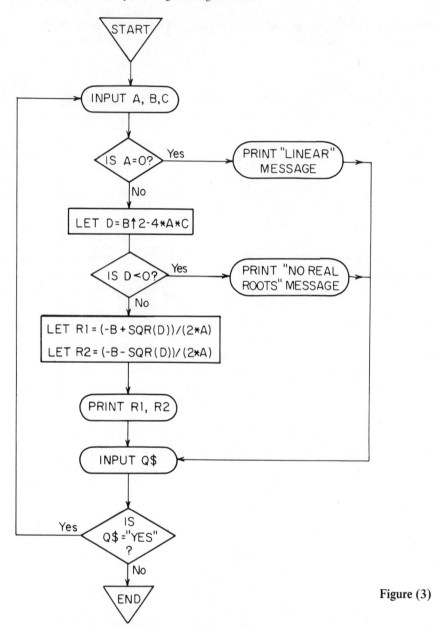

Figure (3)

You will recall that in the quadratic formula as given in Paragraph (2) above the value of A must not be zero. If A is zero, then the original equation is not quadratic but linear. We account for this possibility in the flow chart by including a decision box for the variable A. If A = 0, then the computer should print a message stating that the original equation is linear. The computer should then ask whether we wish to repeat the program.

If A ≠ 0, then the program should continue. The variable D for the discriminant is introduced. We include another decision box in the flow chart to examine the value of D for particular values of A, B, and C. You know from your work in algebra that if D < 0, then there are no real roots for those values of A, B, and C. We then want the computer to print a message to this effect, and then ask whether we wish to repeat the program.

If D >= 0, then we want to define the variables R1 and R2 according to the quadratic formula. Notice that we accounted for the unusual cases first in the flow chart, i.e., when A = 0 and when D < 0.

The computer should then print the values of R1 and R2, and should continue by asking us if we want to execute the program again.

We include a third decision box in the flow chart for the string variable Q$. If Q$ is YES, then we want to design the program so that new values of A, B, and C will be introduced. If Q$ is NO, then we want the program to stop.

As you can see from Figure (3), a flow chart can be a rather detailed and complete description of a program. The flow chart is simply a method by which you can organize your thoughts about a program. Generally speaking, the more time and thought that you spend on a flow chart, the less time you will need to spend on the actual program.

4. Program (4): The Real Roots of $AX^2 + BX + C = 0$

The program for the problem at hand appears on the following page.

5. Comments about Program (4)

a. In Steps 200 and 220, the GO TO command instructs the computer to go to Step 130, thus asking if we should repeat the program.

b. In Step 180, we introduce the STOP command. The STOP command stops the execution of the program. This instruction is equivalent to the command GO TO 230, where Step 230 contains the END statement.

c. Notice the three IF-THEN statements in the program: one in Step 70 concerning A, another in Step 90 about the variable D, and the third in Step 170 for the string variable Q$.

PROGRAM (4):

```
10 PRINT "THIS PROGRAM FINDS THE REAL ROOTS OF ANY QUADRATIC ";
15 PRINT "EQUATION."
20 PRINT
30 PRINT "WHAT ARE THE COEFFICIENTS (START WITH THE X↑2 TERM AND"
40 PRINT "SEPARATE THE COEFFICIENTS WITH COMMAS) ";
50 INPUT A,B,C
60 PRINT
70 IF A=0 THEN 190
80 LET D=B↑2-4*A*C
90 IF D<0 THEN 210
100 LET R1=(-B+SQR(D))/(2*A)
110 LET R2=(-B-SQR(D))/(2*A)
120 PRINT "THE REAL ROOTS OF"A;"X↑2+"B;"X+"C;" = 0 ARE "R1;",";R2
130 PRINT
140 PRINT
150 PRINT "DO YOU WISH TO RUN AGAIN ";
160 INPUT Q$
170 IF Q$="YES" THEN 20
180 STOP
190 PRINT "THE EQUATION IS LINEAR."
200 GOTO 130
210 PRINT "THE EQUATION HAS NO REAL ROOTS."
220 GOTO 130
230 END
```

OUTPUT:

```
THIS PROGRAM FINDS THE REAL ROOTS OF ANY QUADRATIC EQUATION.

WHAT ARE THE COEFFICIENTS (START WITH THE X↑2 TERM AND
SEPARATE THE COEFFICIENTS WITH COMMAS) ?1,2,1

THE REAL ROOTS OF 1 X↑2+ 2 X+ 1 = 0 ARE -1 ,-1


DO YOU WISH TO RUN AGAIN ?YES

WHAT ARE THE COEFFICIENTS (START WITH THE X↑2 TERM AND
SEPARATE THE COEFFICIENTS WITH COMMAS) ?1,0,1

THE EQUATION HAS NO REAL ROOTS.

DO YOU WISH TO RUN AGAIN ?YES

WHAT ARE THE COEFFICIENTS (START WITH THE X↑2 TERM AND
SEPARATE THE COEFFICIENTS WITH COMMAS) ?0,43,2

THE EQUATION IS LINEAR.


DO YOU WISH TO RUN AGAIN ?YES

WHAT ARE THE COEFFICIENTS (START WITH THE X↑2 TERM AND
SEPARATE THE COEFFICIENTS WITH COMMAS) ?25.5,73.431,4.329
```

```
THE REAL ROOTS OF 25.5 X↑2+ 73.431 X+ 4.329 = 0 ARE -6.02123 E-2 ,
-2.81943

DO YOU WISH TO RUN AGAIN ?NO
```

d. Notice that the organization of the program is such that the "linear" message and the "no real roots" message are near the end of the program, followed by the GO TO statement. These two messages could have appeared almost anywhere in the program.

e. Notice in Step 120 the arrangement of quotation marks, semicolons and variables. Study the OUTPUT of that part of the program which relates to Step 120 so that you thoroughly understand this arrangement.

6. The ON-GO TO Command

a. In Section 5-3 you learned how to use an IF-THEN command as a two-way conditional statement. There is a method in BASIC by which we can have a multiple-way conditional switch: the ON-GO TO command.

The general form of the ON-GO TO statement is:

$$\text{ON E GO TO } N_1, N_2, N_3, \ldots$$

where E is any numerical expression or variable, and N_1, N_2, N_3, ... are step numbers.

During execution of an ON-GO TO statement, the computer calculates the integral part of the value for the expression E, which we write INT(E). It then jumps to line N_1 if INT(E) = 1, to line N_2 if INT(E) = 2, to line N_3 if INT(E) = 3, etc. Any number of lines may be designated in an ON-GO TO statement, provided the step numbers N_1, N_2, N_3, ... all fit on one line. If the value of INT(E) is less than 1 or greater than the number of lines specified in N_1, N_2, N_3, ..., then the computer will type an appropriate ERROR message. Note that some systems will not jump to step numbers containing the statements DATA, DEF (see Section 6-4), DIM (see Section 5-8) and REM (see Section 6-1) by means of the ON-GO TO command.

In Example (1) below, the ON-GO TO command in Step 30 sends the computer to Step 40, 60, or 80, depending on the value of X. The value of X is determined from the READ-DATA statements in Steps 10 and 100.

The computer first sets X = 2, since 2 is the first number in the DATA statement. In Step 30, the computer branches to Step 60, because 60 is the second step number in the ON-GO TO command.

After the computer types BRANCH 2 from Step 60, it goes to the next step. The GO TO command of Step 70 sends the computer back to Step 10, and the process is repeated for X = 1.

The computer prints BRANCH 1 from Step 40, executes Step 50 by jumping back to Step 10, and then assigns the value 3.726 to X. Since INT(3.726) = 3, then the computer branches to Step 80 and prints BRANCH 3.

The computer executes Step 90, jumps back to Step 10, and reads the next value of X = -100. Since the condition of Step 20 is now satisfied, i.e., X = -100, the computer jumps to Step 110 and the program has been completely executed.

We call the number -100 a *flag*. We used -100 as the flag because in an ON-GO TO statement, we cannot have a negative number for the value of X.

EXAMPLE (1)

PROGRAM:

```
10 READ X
20 IF X=-100 THEN 110
30 ON X GOTO 40,60,80
40 PRINT "BRANCH 1"
50 GOTO 10
60 PRINT "BRANCH 2"
70 GOTO 10
80 PRINT "BRANCH 3"
90 GOTO 10
100 DATA 2,1,3.726,-100
110 END
```

OUTPUT:

```
BRANCH 2
BRANCH 1
BRANCH 3
```

b. The ON-GO TO command is frequently used in conjunction with the function SGN(X). The statement ON SGN(E)+2 GO TO N_1, N_2, N_3 directs the computer to jump to line N_1 if $E < 0$, to line N_2 if $E = 0$, and to line N_3 if $E > 0$.

c. Some systems use the following format for the ON-GO TO statement:

$$\text{GO TO } (N_1, N_2, N_3, \ldots) \, E$$

where N_1, N_2, N_3, ... are line numbers, and E is a numerical expression or a variable. The computer evaluates INT(E) and branches to line N_1 if INT(E) = 1, to N_2 if INT(E) = 2, to N_3 if INT(E) = 3, etc.

On some systems, the command ON E THEN N_1, N_2, N_3, ... may be used in place of ON E GO TO N_1, N_2, N_3, ... The two commands are equivalent.

EXERCISES 5-5

Follow directions (a) through (g) in Exercises 5-2 for each of the following problems. Provide for repetition by using a string variable.

1. Write a program which asks for the constants A, B, C, and D in the equation AX+B = CX+D, and then types the solution.
2. Write a program which asks for the constants A1, A2, A3, and B1, B2, B3 in the set of simultaneous equations

$$A1X+A2Y = A3$$
$$B1X+B2Y = B3$$

and then types the solutions. Consider inconsistent and dependent cases, in addition to the unique solution case.
3. Do problem (2) above for three equations in three variables. Consider *all* cases.
4. Ask for three numbers A, B, and C and then type the numbers in order of magnitude, from largest to smallest.
5. Ask for three words A$, B$, and C$ and then type the words in alphabetical order.
6. Ask for two sets of ordered pairs (X(1), Y(1)), and (X(2), Y(2)), and then type the equation of a straight-line graph which passes through the points whose ordered pairs are given. Have the answer typed out in the form of an equation.
7. Write a program which asks for a number N, and then tells which of the following properties that number has:

        a) integral   d) negative
        b) nonintegral   e) zero
        c) positive

5-6 Loops; Program (5): N! (Factorial)

1. The Loop

We shall now discuss one of the most powerful uses of the computer: the loop. A loop is simply an iterative process that continues until a given condition is satisfied. If no condition is satisfied, then the loop will continue until the limits of the computer are reached, or until you escape from the loop by typing a character which controls the system. (Some computers recognize an endless loop and stop automatically.)

Look at Example (1) below. Step 10 *initializes* the value of I by setting I = 0. The loop in Steps 20 through 40 reads 11 values and stores them respectively in A(0) through A(10). That is, Step 20 first reads the value for A(0), namely, 1. Then in Step 30 I is incremented by 1. Step 40 determines whether or not $I < 11$. Since $1 < 11$, the computer jumps back to Step 20 and reads A(1). The variable I is again incremented by 1, so that I = 1+1 = 2. Since $2 < 11$, the computer again jumps back to Step 20 and reads A(2). This loop is repeated until the computer reads the value of A(10). In Step 30, during this 11th cycle of the loop, I = 11. Since the conditional IF of Step 40 is not satisfied, the computer continues to Step 50 and thus prints the values of A(0) and A(10), as you can see in the OUTPUT.

Initializing I = 0 in Step 10 is not absolutely necessary in Example (1), because the computer assigns the value 0 to all undefined variables. It is good programming practice to set all such *counters* equal to 0, however, because in a program which provides for repetition the initial value of the variable on the second RUN (during the same ON LINE session) would be the value of that variable at the end of the first RUN, and that would not necessarily be 0.

EXAMPLE (1)

PROGRAM:

```
10 LET I=0
20 READ A(I)
30 LET I=I+1
40 IF I<11 THEN 20
50 PRINT A(0),A(10)
60 DATA 1,3,5,7,9,11,13,15,17,19,21
70 END
```

OUTPUT:

```
1              21
```

2. N! (Factorial)

A more significant program which utilizes the idea of a loop is a program which asks for a nonnegative integer N, and then computes and types N! (factorial), where N! is defined as follows:

$$N! = N(N-1)(N-2) \ldots (3)(2)(1).$$

If N = 0, then we define: 0! = 1.

Let us draw a flow chart and then write a program which will compute N!.

3. The Flow Chart for N! (Factorial)

The flow chart for N! appears on the following page.

Study the flow chart carefully. Since N! is defined only for nonnegative integers, two decision boxes must be included in the flow chart to examine N. If N is less than zero, or if N is not an integer, the computer should print a message telling us so, and then ask for another value of N.

F = F*X is probably the most important statement in the flow chart and the subsequent program. It enables the computer to accumulate the value of N!. First, F = 1 and X = 1. Then F = F * X. On the first loop of the intended program, F = 1*1. X is then incremented by 1 and its value is examined in the decision box to determine whether or not X is less than or equal to N. If $X < = N$, then the computer goes through the loop a second time.

The computer again calculates the value of F. Since X is now equal to 2, F = 1*2 = 2. X is again incremented by 1 and its value is compared with N. If $X \leqslant N$, the computer goes through the loop a third time, with X equal to 3. The statement F = F*X yields F = 2*3 = 6 on this third loop.

The computer continues to execute the loop until $X > N$. The computer then types the last stored value of F, which is N!, along with the value of N.

The last decision box in the flow chart is concerned with repetition of the program. If Q$ is YES, the computer should ask for another value of N, and the entire procedure for calculating and printing N! should be repeated.

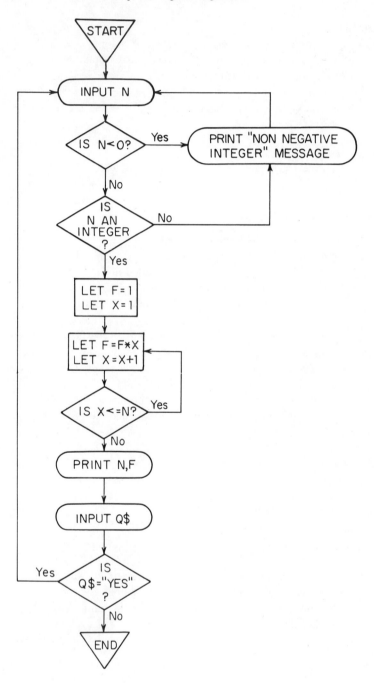

Figure (4)

4. Program (5): N! (Factorial)

The program of the above flow chart to compute N! is as follows.

PROGRAM (5)

```
10 PRINT "THIS PROGRAM COMPUTES N FACTORIAL FOR ANY NONNEGATIVE ";
20 PRINT "INTEGER N."
30 PRINT
40 PRINT "WHAT IS N ";
50 INPUT N
60 PRINT
70 IF N<0 THEN 190
80 IF N>INT(N) THEN 190
90 LET F=X=1
100 LET F=F*X
110 LET X=X+1
120 IF X<=N THEN 100
130 PRINT N;"FACTORIAL IS"F
140 PRINT
150 PRINT "DO YOU WISH TO RUN AGAIN ";
160 INPUT Q$
170 IF Q$="YES" THEN 30
180 STOP
190 PRINT "N MUST BE A NONNEGATIVE INTEGER."
200 GOTO 30
210 END
```

OUTPUT:

```
THIS PROGRAM COMPUTES N FACTORIAL FOR ANY NONNEGATIVE INTEGER N.

WHAT IS N ?3

 3 FACTORIAL IS 6

DO YOU WISH TO RUN AGAIN ?YES

WHAT IS N ?5

 5 FACTORIAL IS 120

DO YOU WISH TO RUN AGAIN ?YES

WHAT IS N ?4.3

N MUST BE A NONNEGATIVE INTEGER.

WHAT IS N ?33

 33 FACTORIAL IS 8.68332 E+36

DO YOU WISH TO RUN AGAIN ?YES

WHAT IS N ?0

 0 FACTORIAL IS 1

DO YOU WISH TO RUN AGAIN ?NO
```

5. Comments about the Program (5)

a. Look at Step 90 of the above program. Notice that we must first assign a value to the variable F and X. Otherwise the computer would assume that they are 0, and N factorial would always equal 0.

b. As stated in Paragraph 3 above, the statement F = F*X in Step 100 is probably the most important step in the entire program; it is the step by which the computer can work with the definition of N!. Notice that the machine is computing N! = (1)(2)(3) ... (N-2)(N-1)(N) by accumulating the product in F = F*X with X=1, X=2,..., X=N. In Step 100, the value of F on the right-hand side of the equal sign is the value from the previous loop, and the value of F on the left-hand side of the equal sign is the value in the current loop. When $X > N$, the computer stops looping and in Step 130 types F, which is the value of N!.

c. In Step 80 we ensure that N is an integer; in Step 70 the computer examines the value of N to determine if it is a negative number. If N is not a positive integer or zero, then, as designed in our flow chart, the computer jumps to Step 190, prints the message N MUST BE A NONNEGATIVE INTEGER, and then asks for a new value of N in Step 50.

d. Notice in the last part of the OUTPUT that the program correctly computes 0! = 1. The reason for this is as follows. In Step 90 F and X are set equal to 1. Then in Step 100 F is redefined as 1*1, or 1. X is then incremented by 1. Since $X > N$, the condition in Step 120 is not satisfied, and the computer proceeds to type the value of F = 1 for 0!.

e. Because a loop is so frequently used in programming, BASIC contains a reiterative procedure which makes the writing of a program with a loop much simpler than our Program (5). This procedure is illustrated in Section 5-7.

EXERCISES 5-6

Follow directions (a) through (g) of Exercises 5-2 for each of the following problems. Provide for repetition by using string variables.

1. Write a program which asks for three integers A, B, and C, and then types the greatest common divisor of those numbers.

2. Do Problem (1) above for the least common multiple of three numbers A, B, and C.

3. Write a program which asks for a positive integer and then tells whether or not that integer is prime.

4. Write a program which asks for a positive integer and then finds all the prime factors of that integer.

5. Write a program which asks for a positive integer N and then finds all pairs of factors of N.

5-7 The FOR-NEXT Command; Program (6): N!, A Second Approach

1. Constructing a Loop with the FOR-NEXT Command

Because loops are used so frequently in programming, the BASIC language is designed to make looping quite simple. A loop can be easily constructed by the following commands:

$$\text{FOR } V = E_1 \text{ TO } E_2 \text{ STEP } E_3$$
$$\cdot$$
$$\cdot$$
$$\cdot$$
$$\text{NEXT } V$$

V is any nonsubscripted variable, and E_1, E_2 and E_3 are numerical expressions or variables. E_1 is the initial value of the variable V, E_2 is the final value, and E_3 is the increment value.

The FOR statement appears at the beginning of the loop, and the NEXT statement is written at the end of the loop.

Look at Example (1). In the loop contained in Steps 10 through 30, the computer sets the values of $I(1) = 1\uparrow 2+1 = 2$, $I(2) = 2\uparrow 2+1 = 5$, $I(3) = 3\uparrow 2+1 = 10, \ldots, I(10) = 10\uparrow 2+1 = 101$. That is, the initial value of the control variable X is set equal to 1 in Step 10. (The initial value of the control variable appears between the equal sign and the word TO in the FOR statement.) Then, in Step 20, I(X) or I(1) is set equal to $X\uparrow 2+1=1\uparrow 2+1=2$. The NEXT command in Step 30 instructs the computer to increment the value of X by the value of the number or expression appearing after the word STEP in the FOR statement. Thus, since we are incrementing by 1 in this example, X is set equal to 2.

X is now compared with the final value (10) of the control variable. (The final value appears between the words TO and STEP in the FOR statement.) Since $X \leq 10$, the computer goes through the loop a second time. The machine sets I(X) or I(2) equal to $X\uparrow 2+1 = 2\uparrow 2+1 = 5$ and again increments X by 1. This new value of X (3) is again compared with the final value (10) of the control variable. Since $X \leq 10$, the computer executes the loop a third time.

This process continues until X = 10. The computer then sets I(X) or I(10) equal to $10\uparrow 2+1 = 101$ and, in Step 30, increments X by 1. Since X is now greater than 10, the computer stops executing the loop. The value of X is then automatically *decreased* by the increment value 1. Thus, X is again set equal to 10. The computer proceeds to execute Step 40 by printing the values of I(1) and I(X) or I(10), as you can see in the OUTPUT.

It is good programming practice to indent the statements inside a loop, as illustrated in Step 20. This practice permits us easily to see a loop in a program.

EXAMPLE (1)

PROGRAM:

```
10 FOR X=1 TO 10 STEP 1
20    LET I(X)=X↑2+1
30 NEXT X
40 PRINT I(1), I(X)
50 END
```

OUTPUT:

```
 2                 101
```

If we omit Step 30 NEXT X from Example (1), we get an appropriate ERROR message, as shown in Example (2) below.

EXAMPLE (2)

PROGRAM:

```
10 FOR X=1 TO 10 STEP 1
20    LET I(X)=X↑2+1
40 PRINT I(1), I(X)
50 END
```

OUTPUT:

```
FOR WITHOUT NEXT IN 10
```

Thus, we must always have a NEXT statement to accompany each FOR statement in a program. Furthermore, the variable name appearing to the left of the equal sign in the FOR statement must always be the same as the variable name in the NEXT statement.

If we omit the increment value and the word STEP from the FOR statement, as in Step 20 of Example (3) below, the computer assumes that each value of X should be incremented by 1.

EXAMPLE (3)

PROGRAM:

```
10 PRINT "NUMBER", "SQUARE"
20 FOR X=1 TO 5
30    PRINT X,X↑2
40 NEXT X
50 END
```

OUTPUT:

NUMBER	SQUARE
1	1
2	4
3	9
4	16
5	25

The initial value of the FOR command, the increment value, and the final value may be any number or a well-defined expression. Thus, in Example (4) the initial value is .1, the increment value is .1, and the terminal value is .5.

EXAMPLE (4)

PROGRAM:

```
10 PRINT "NUMBER", "SINE"
20 FOR X=.1 TO .5 STEP .1
30    PRINT X, SIN(X)
40 NEXT X
50 END
```

OUTPUT:

```
NUMBER         SINE
  .1           9.98334 E-2
  .2            .198669
  .3            .295520
  .4            .389418
  .5            .479426
```

In Example (5) the initial value is .5, the increment value is (-.1), and the terminal value is 0. Of course, in Example (4) the increment value is positive, because the final value is larger than the initial value. In Example (5) the increment value is negative, because the final value is smaller than the initial value.

EXAMPLE (5)

PROGRAM:

```
10 PRINT "NUMBER", "SINE"
20 FOR X=0.5 TO 0 STEP -0.1
30    PRINT X, SIN(X)
40 NEXT X
50 END
```

OUTPUT:

```
NUMBER         SINE
  .5            .479426
  .4            .389418
  .3            .295520
  .2            .198669
  .1           9.98334 E-2
 5.58794 E-9   5.58794 E-9
```

Example (6) is a program which illustrates that the initial value, the increment value, and the final value may be a well-defined expression.

EXAMPLE (6)

PROGRAM:

```
10 PRINT "TYPE THREE NUMBERS ";
20 INPUT A,B,C
30 PRINT "NUMBER", "LOGARITHM"
40 FOR X=A+1 TO C+2 STEP A+B
50     PRINT X, LOG(X)
60 NEXT X
70 END
```

OUTPUT:

```
TYPE THREE NUMBERS ?0.1,0.4,1.6
NUMBER          LOGARITHM
  1.1            9.53102 E-2
  1.6            .470004
  2.1            .741937
  2.6            .955511
  3.1           1.13140
  3.6           1.28093
```

If the increment value is positive, the loop continues to be executed as long as the value of the control variable is less than or equal to the final value, as in Example (7).

EXAMPLE (7)

PROGRAM:

```
10 K=0
20 FOR I=1 TO 2 STEP .6
30     K=K+1
40 NEXT I
50 PRINT "THE LOOP WAS EXECUTED"K;"TIMES."
60 END
```

OUTPUT:

```
THE LOOP WAS EXECUTED 2 TIMES.
```

If the increment value is negative, the loop continues to be executed as long as the value of the control variable is greater than or equal to the final value, as in Example (8).

If the initial and final values are equal, then the loop is executed once. If the initial value is greater than the final value for positive increment numbers, or if the initial value is less than the final value for negative increment numbers, then the loop is not performed at all.

EXAMPLE (8)

PROGRAM:

```
10 K=0
20 FOR I=3 TO 1 STEP -0.7
30    K=K+1
40 NEXT I
50 PRINT "THE LOOP WAS EXECUTED"K;"TIMES."
60 END
```

OUTPUT:

```
THE LOOP WAS EXECUTED 3 TIMES.
```

2. Nested Loops

Example (9) shows that loops may be placed inside other loops. Such arrangements of loops are said to be *nested*.

EXAMPLE (9)

PROGRAM:

```
10 PRINT " A", " B", "A+B"
20 FOR A=1 TO 4
30    FOR B=1 TO 3
40       PRINT A,B,A+B
50    NEXT B
60 NEXT A
70 END
```

OUTPUT:

A	B	A+B
1	1	2
1	2	3
1	3	4
2	1	3
2	2	4
2	3	5
3	1	4
3	2	5
3	3	6
4	1	5
4	2	6
4	3	7

5-7 The FOR-NEXT Command; Program (6)

Notice in Example (9) above that the inner loop increments faster than the outer loop. That is, when A=1, B=1, 2,3; when A=2, B=1, 2, 3; etc.

Loops may be nested to great depths, but they may not cross one another, as the following diagram illustrates.

Table (3). NESTED LOOPS

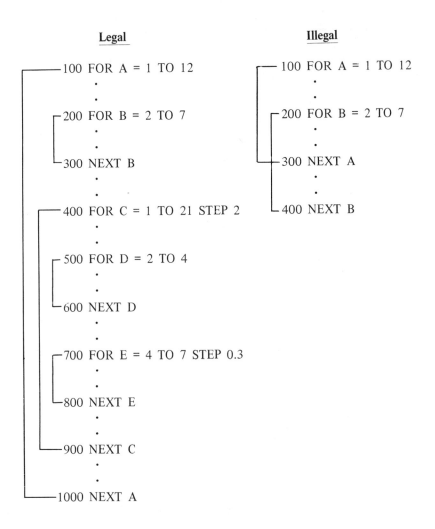

3. The Transfer of Control Outside the Loop

Control of the program may be transferred from within a loop to steps outside the loop, as Example (10) below illustrates. The computer goes through the loop until J < 0.1 or I = 10, whichever occurs first. You see in the OUTPUT of Step 60 that the condition J < 0.1 held true when I = 4. The condition J < 0.1 occurred before the condition I = 10, and thus the statement of Step 40 (inside the loop of Steps 20 through 50) directed the computer outside the loop to Step 60.

EXAMPLE (10)

PROGRAM:

```
10 J=1
20 FOR I=1 TO 10
30    J=J/2
40    IF J<0.1 THEN 60
50 NEXT I
60 PRINT I,J
70 END
```

OUTPUT:

```
4            .0625
```

4. The Flow Chart for Program (6): N!, A Second Approach

We are now ready to write another version of Program (5), using a FOR-NEXT loop. We first construct the flow chart.

The new aspect of the flow chart of Figure (5) from the one shown in Figure (4) is the diamond-shaped symbol containing the statement FOR X=2 TO N. We use this diamond symbol to portray a loop in the resulting program. One arrow from the diamond goes to the contents of the loop (LET F = F*X). A second arrow enters the diamond from the end of the loop. A third arrow from the loop goes to that part of the program which follows the loop. We label this arrow DONE to indicate that the computer should follow this path in the program after completely executing the loop.

5. Program (6): N!, A Second Approach

The program for computing N! using the FOR-NEXT statements is shown on page 100.

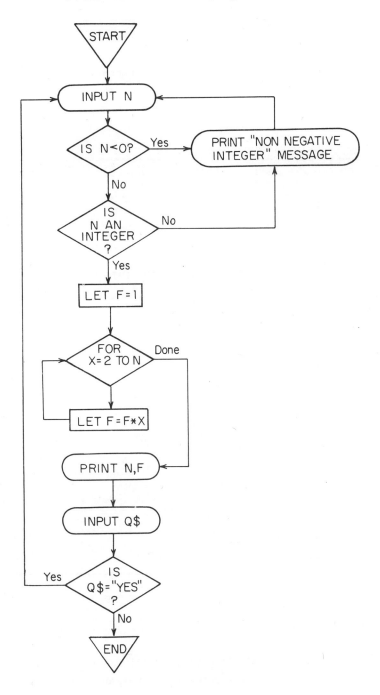

Figure (5)

PROGRAM (6)

```
 10 PRINT "THIS PROGRAM COMPUTES N FACTORIAL FOR ANY NONNEGATIVE ";
 20 PRINT "INTEGER N."
 30 PRINT
 40 PRINT "WHAT IS N ";
 50 INPUT N
 60 PRINT
 70 IF N<0 THEN 190
 80 IF N>INT(N) THEN 190
 90 LET F=1
100 FOR X=2 TO N
110     LET F=F*X
120 NEXT X
130 PRINT N;"FACTORIAL IS"F
140 PRINT
150 PRINT "DO YOU WISH TO RUN AGAIN ";
160 INPUT Q$
170 IF Q$="YES" THEN 30
180 STOP
190 PRINT "N MUST BE A NONNEGATIVE INTEGER."
200 GOTO 30
210 END
```

OUTPUT:

```
THIS PROGRAM COMPUTES N FACTORIAL FOR ANY NONNEGATIVE INTEGER N.
WHAT IS N ?4

 4 FACTORIAL IS 24

DO YOU WISH TO RUN AGAIN ?YES

WHAT IS N ?-10

N MUST BE A NONNEGATIVE INTEGER.

WHAT IS N ?12

 12 FACTORIAL IS 4.79002 E+8

DO YOU WISH TO RUN AGAIN ?YES

WHAT IS N ?0

 0 FACTORIAL IS 1

DO YOU WISH TO RUN AGAIN ?NO
```

6. Comments About Program (6)

a. The significant aspect of Program (6) is contained in Steps 90 through 120. Step 90 sets the initial value of F=1. The loop using the

FOR-NEXT command is contained in Steps 100 through 120. The computer first goes through the loop with X=2: F=1*2 = 2. In the second loop, X=3 and F=2*3 = 6. In the third loop, X=4 and F=6*4 = 24. The computer goes through the loop N-1 times (unless N = 0). After the last loop, the computer continues to execute the program from Step 130 through Step 170.

b. Notice that in Step 110, unless X=2, the value of F on the right side of the equal sign is the value from the previous loop, and the value of F on the left side of the equal sign is the value in the current loop.

EXERCISES 5-7

1. Write a program which asks for N positive integers, then finds the least common multiple of those N numbers. Provide for repetition.
2. Write a program which types all the Pythagorean triplets up to a given positive integer N. Provide for repetition.
3. Write a program which calculates the ratio $SIN(X)/X$ as X goes from 1 to 1E-6. What number is this ratio approaching?
4. Write a program which asks for a number N, and then types all the prime numbers less than N. Provide for repetition.
5. Write a program which asks for any number N, and then calculates the square root of N using the formula $S_{i+1} = (N/S_i + S_i)/2$, where S_i is the *ith* approximation and S_{i+1} is the $(i+1)st$ approximation. Let the first approximation be $S_1 = N/2$. The program should ask for the number of approximations A. (Do not use subscripts, i.e., leave out the subscripts of the above formula in your program.)
6. Write a program which calculates $A = (1 + X) \uparrow (1/X)$ as X goes from 1 to 1E-6. What number is A approaching?

5-8 The Command DIM

1. Its Use with Subscripted Variables

A subscripted variable is very useful for handling a large block of data.

For example, the use of subscripted variables is an effective method of assigning values to many variables when used in conjunction with the FOR-NEXT command. You learned in Section 5-7 that the FOR-NEXT loop given by

```
          ⋮
110  FOR I = 0 TO 10
120     READ A(I)
130  NEXT I
140  DATA 11,10,9,8,7,6,5,4,3,2,1
          ⋮
```

will read 11 values and assign them to the storage areas which we call A(0), A(1), A(2), ..., A(10).

When using subscripts greater than 10, however, we must inform the computer to reserve enough storage locations for these variables. We reserve space for these subscripted variables with the command DIM. Consider the following example.

EXAMPLE (1)

PROGRAM:

```
10 FOR I=1 TO 11
20    READ A(I)
30 NEXT I
40 PRINT A(11)
50 DATA 20,19,18,17,16,15,14,13,12,11,10
60 END
```

OUTPUT:

```
DIMENSION ERROR IN 20
```

The ERROR message in Example (1) above tells us that a DIM statement is required in the program. The DIM statement is necessary because the variables in Step 20 have subscripts greater than 10. Let us insert a DIM statement in Example (1).

EXAMPLE (2)

PROGRAM:

```
 5 DIM A(15)
10 FOR I=1 TO 11
20     READ A(I)
30 NEXT I
40 PRINT A(11)
50 DATA 20,19,18,17,16,15,14,13,12,11,10
60 END
```

OUTPUT:

10

In Step 5 of Example (2) above, we told the computer to reserve 16 locations for the variable A, from A(0) to A(15). We may reserve as much space as we want. However, we should keep in mind that the computer has a limited total storage space for each user. A program with very large DIM statements must necessarily be a short program. We should reserve only as much space as we need.

2. More Than One Variable

There may be more than one variable in a DIM statement, as in Example (3) below. If no DIM statement appears for a variable, then the computer automatically reserves space for subscripts less than or equal to 10. That is why you see no DIM statement for the variable C in Example (3).

3. A DIM Statement before a Variable

Some systems require that the DIM statement precede the use of subscripted variables in a program, as in Example (4) below. On other systems they can be placed anywhere in the program. Most programmers put DIM statements at the beginning of a program.

EXAMPLE (3)

PROGRAM:

```
 10 DIM A(15),B(21)
 20 FOR I=1 TO 15
 30    READ A(I)
 40 NEXT I
 50 RESTORE
 60 FOR I=1 TO 21
 70    READ B(I)
 80 NEXT I
 90 RESTORE
100 FOR I=0 TO 10
110    READ C(I)
120 NEXT I
130 PRINT A(2),B(7),C(4)
140 DATA 1,2,3,4,5,6,7,8,9,10,11,12,13,14,15,16,17,18,19,20,21
150 END
```

OUTPUT:

```
 2              7              5
```

EXAMPLE (4)

PROGRAM:

```
10 DIM A(20)
20 FOR I=5 TO 20
30    READ A(I)
40 NEXT I
50 PRINT A(11)
60 DATA 10001,10002,10003,10004,10005,10006,10007,10008
70 DATA 10009,10010,10011,10012,10013,10014,10015,10016
80 END
```

OUTPUT:

```
10007
```

4. Subscripted String Variables

String variables using subscripts greater than ten must also be DIMensioned, just as numerical variables. Thus you see a DIMension ERROR in Example (5), but after including a DIM statement in Step 5 of Example (6), the program ran properly.

EXAMPLE (5)

PROGRAM:

```
10 FOR I=1 TO 12
20    READ A$(I)
30 NEXT I
40 PRINT A$(12)
50 DATA A,B,C,D,E,F,G,H,I,J,K,L
60 END
```

OUTPUT:

DIMENSION ERROR IN 20

EXAMPLE (6)

PROGRAM:

```
 5 DIM A$(12)
10 FOR I=1 TO 12
20    READ A$(I)
30 NEXT I
40 PRINT A$(12)
50 DATA A,B,C,D,E,F,G,H,I,J,K,L
60 END
```

OUTPUT:

L

EXERCISES 5-8

Follow the directions of Exercises 3-1 for each of the following problems:

1.
```
10 FOR I = 1 TO 11
20    READ A(I)
30 NEXT I
40 PRINT A(11)
50 DATA 1,2,3,4,5,6,7,8,9,10,11
60 END
```

2.
```
 5  DIM A(15)
10  FOR I = 1 TO 11
20     READ A(I)
30  NEXT I
40  PRINT A(11)
50  1,2,3,4,5,6,7,8,9,10,11
60  END
```

3.
```
10  FOR I = 1 TO 10
20     READ B$(I)
30  NEXT I
40  PRINT B$(3)
50  DATA P,Q,R,S,T,U,V,W,X,Y
60  END
```

5-9 Summary

You began to appreciate the real power of the computer in this chapter. You learned how to draw a flow chart, and saw the ease with which we can write a program from a carefully designed flow chart.

After you execute a program, you now know what to do with it: you can file the program onto the disc storage unit, and you can make a LIST of the program. At the same time, you can make another tape of the program. You also know how to retrieve a file from the disc.

You learned about the IF-THEN command and saw it used in a number of different contexts throughout the chapter. You now realize that it is not necessary to compile a program each time you want to execute it; you can use the IF-THEN statement to RUN the program again.

You also learned about the alphanumeric capabilities of the BASIC language, and how to use string variables. You learned that the GO TO command directs the computer to another step in the program, from which it continues to execute the program. You also saw a few examples of the ON-GO TO command as a method of multiple-way switching to other steps in a program.

You discovered what a loop in a program is, and how easily you can write a loop using the FOR-NEXT commands. You learned that subscripted variables with subscripts greater than 10 require the DIM statement to reserve space for the variables.

CHAPTER 6

In this chapter, you will learn how to write a subroutine in a program. You will also learn how to input an unspecified number of data entries with the command MAT INPUT, and that the computer automatically sets NUM equal to the number of data entries which you typed.

You will become familiar with the CHANGE command — that command which converts each character of a string into a numerical code number. Finally, you will learn how to define your own functions in a program.

All these ideas are developed within the context of particular programs, and in each program you will need to draw upon your programming knowledge and experience of Chapters 1-5.

6-1 The GOSUB-RETURN Commands; Program (7): A Dice Game

1. Subroutine

A useful programming technique is to direct the computer to another part of a program, called a *subroutine,* instruct the computer to execute that part, and then tell the computer to return to the main section of the program. This technique is particularly useful for those programs which require a set of instructions which must be repeated several times, but at different places in the program.

2. Commands GOSUB-RETURN

The command which directs the computer to a subroutine is GOSUB. The general form of the command is

GOSUB N ,

where N is a line number in the program. On some systems, N may not be a nonexecutable statement (DATA, DEF [See Section 6-4], DIM or REM [See Paragraph 4 in this Section]). The GOSUB command directs the computer to execute a subroutine beginning at Step N. The computer continues to execute the statements following Step N until it encounters a RETURN command. The RETURN command sends the computer back to the step following the GOSUB instruction.

Look at Program (7). It simulates a game of dice, for which the directions are given in Steps 130 through 230. This program is included here primarily to illustrate the GOSUB-RETURN commands, but there are included several other important programming techniques which will be discussed below.

The GOSUB command is first given in Step 400. When the computer reaches this step during execution, it is directed to the subroutine which begins in Step 710. The subroutine of Steps 710 through 750 simulates the rolling of two fair dice by means of the random function RND, assigns the number of these "rolls" to the variables R1 and R2, finds their sum R, and then prints the values R1 and R2. The RETURN command in Step 750 directs the computer to Step 410, the step following the GOSUB command in Step 400.

PROGRAM (7):

```
100 REM WE ARE NOT RESPONSIBLE FOR THE USE OF THIS OR ANY SIMILAR
110 REM PROGRAMS.
120 REM-----------DIRECTIONS------------
130 PRINT "   THIS PROGRAM SIMULATES A DICE GAME."
140 PRINT "   THE COMPUTER FIRST ASKS YOU HOW MUCH YOU WANT TO BET."
150 PRINT "THE HOUSE LIMIT IS $1000."
160 PRINT "   THE MACHINE THEN ROLLS THE DICE.  IF THE SUM OF THE"
170 PRINT "DICE IS 7 OR 11, THE COMPUTER WINS.  IF THE SUM IS 2, 3,"
180 PRINT "OR 12, YOU WIN."
190 PRINT "   IF ANY OTHER COMBINATION TURNS UP, THE COMPUTER AGAIN"
200 PRINT "CASTS THE DICE.  IF THE SUM IS 7, YOU WIN.  IF THE SUM OF"
210 PRINT "THE DICE IS THE SAME AS THAT OF THE FIRST ROLL, THE COMPUTER"
220 PRINT "WINS.  THE COMPUTER CONTINUES TO ROLL THE DICE UNTIL ONE OF"
230 PRINT "THESE CONDITIONS IS SATISFIED."
240 RANDOM
250 REM-----------PLACING BET------------
260 T=0       'INITIALIZE COUNTER OF WINNINGS AND LOSES
270 PRINT
280 PRINT
290 PRINT
300 PRINT "WHAT IS YOUR BET ";
310 INPUT B
320 B=INT(B*100)*0.01      'ROUND OFF B TO NEAREST CENT
330 IF B<=1000 THEN 360
340 PRINT "I CAN'T ACCEPT BETS OVER $1000."
350 GO TO 290
360 PRINT
370 REM-----------PLAYING GAME------------
```

6-1 The GOSUB-RETURN Commands; Program (7)

```
380 PRINT " ","FIRST DIE","SECOND DIE"
390 PRINT "FIRST ROLL:",
400 GOSUB 710
410 IF (R-7)*(R-11)=0 THEN 490        'COMPUTER WINS
420 IF (R-2)*(R-3)*(R-12)=0 THEN 530   'COMPUTER LOSES
430 F=R
440 PRINT "NEXT ROLLS:",
450 GOSUB 710
460 IF R=7 THEN 530       'COMPUTER LOSES
470 IF R<>F THEN 450      'IF R=F, COMPUTER WINS. IF R<>F, ROLL AGAIN.
480 REM------------COMPUTER WINS-----------
490 PRINT "SORRY, YOU LOST.   ";
500 T=T-B       'DECREASE TOTAL BY AMOUNT OF BET
510 GOTO 560
520 REM------------COMPUTER LOSES------------
530 PRINT "YOU WON.   ";
540 T=T+B       'INCREASE TOTAL BY AMOUNT OF BET
550 REM------------TOTAL OF BETS------------
560 ON SGN(T)+2 GOTO 570,590,610     'BRANCH DEPENDING ON SIGN OF TOTAL
570 PRINT "YOU HAVE NOW LOST $"ABS(T)
580 GOTO 630
590 PRINT "NOW WE'RE EVEN."
600 GOTO 630
610 PRINT "YOUR TOTAL WINNINGS ARE $"T
620 REM------------REPEAT ?------------
630 PRINT
640 PRINT
650 PRINT
660 PRINT "DO YOU WANT TO PLAY AGAIN ";
670 INPUT Q$
680 IF Q$>="Y" THEN 270
690 STOP
700 REM------------SUBROUTINE WHICH ROLLS DICE------------
710 R1=INT(6*RND(X)+1)
720 R2=INT(6*RND(X)+1)
730 R=R1+R2
740 PRINT TAB(18);R1;TAB(33);R2
750 RETURN
760 END
```

OUTPUT:

```
    THIS PROGRAM SIMULATES A DICE GAME.
    THE COMPUTER FIRST ASKS YOU HOW MUCH YOU WANT TO BET.
THE HOUSE LIMIT IS $1000.
    THE MACHINE THEN ROLLS THE DICE.  IF THE SUM OF THE
DICE IS 7 OR 11, THE COMPUTER WINS.  IF THE SUM IS 2, 3,
OR 12, YOU WIN.
    IF ANY OTHER COMBINATION TURNS UP, THE COMPUTER AGAIN
CASTS THE DICE.  IF THE SUM IS 7, YOU WIN.  IF THE SUM OF
THE DICE IS THE SAME AS THAT OF THE FIRST ROLL, THE COMPUTER
WINS.  THE COMPUTER CONTINUES TO ROLL THE DICE UNTIL ONE OF
THESE CONDITIONS IS SATISFIED.


WHAT IS YOUR BET ?100
                FIRST DIE      SECOND DIE
FIRST ROLL:         3              4
SORRY, YOU LOST.  YOU HAVE NOW LOST $ 100
```

```
DO YOU WANT TO PLAY AGAIN ?YES


WHAT IS YOUR BET ?100
                FIRST DIE       SECOND DIE
FIRST ROLL:        1                2
YOU WON.  NOW WE'RE EVEN.


DO YOU WANT TO PLAY AGAIN ?YES


WHAT IS YOUR BET ?500
                FIRST DIE       SECOND DIE
FIRST ROLL:        5                2
SORRY, YOU LOST.  YOU HAVE NOW LOST $ 500


DO YOU WANT TO PLAY AGAIN ?YES


WHAT IS YOUR BET ?1000
                FIRST DIE       SECOND DIE
FIRST ROLL:        1                4
NEXT ROLLS:        6                6
                   2                4
                   2                2
                   4                5
                   2                5
YOU WON.  YOUR TOTAL WINNINGS ARE $ 500


DO YOU WANT TO PLAY AGAIN ?NO
```

3. The Distinction Between GOSUB and GO TO Commands

Notice the distinction between the GO TO command of Step 350 and the GOSUB command of Step 400 in Program (7). The GO TO command directs the computer to another step in the program (Step 290), from which it does not return unless directed by another GO TO, an IF-THEN or an ON-GO TO command. The GOSUB command directs the computer to another step in the progam (Step 710). After the computer has executed the subroutine (Steps 710 through 740), the RETURN command of Step 750 sends it back to the step following the GOSUB command (Step 410).

4. Comments about Program (7)

a. Notice the use of REM statements and apostrophes throughout the program. The REM statement allows us to insert explanatory remarks in a program. BASIC ignores everything typed to the right of the word REM. Notice in the REM statements of Steps 120, 250, 370, 480, 520, 550, 620, and 700 that we have used dashes and phrases to set off and highlight various parts of the program.

A second method of inserting remarks, one available on most systems, is to use the apostrophe. The apostrophe is placed at the end of a statement, followed by a comment, as in Step 260. The computer ignores the comment during the execution of the program.

b. Notice the statements which are used to place a bet in Steps 260 through 360. If the value of the bet B is greater than 1000, then the computer responds with the message I CAN'T ACCEPT BETS OVER $1000 (Step 340) and then asks for another bet by the GO TO command of Step 350.

c. Notice the technique used to determine whether or not the value of R (the sum of the numbers on the "dice") is equal to 7 or 11 in Step 410. If $R = 7$ or $R = 11$, then $(R-7) = 0$ or $(R-11) = 0$. Therefore, $(R-7)*(R-11) = 0$ and the condition for the computer to win the bet on the first "throw" of the dice is then satisfied. Similarly, if $R = 2$, $R = 3$, or $R = 12$, then the computer loses the bet on the first throw (Step 420).

d. If neither of the conditions of Steps 410 and 420 is satisfied, the computer continues to execute the program. In Step 450, the program jumps to the subroutine beginning in Step 710 a second time. If the next value of R is 7, then according to the rules of the dice game, the computer loses the bet, the program jumps to Step 530, and the computer then prints the message YOU WON.

If $R \neq 7$, the computer continues to Step 470 and compares the value of R and F, where F is the value of the sum of the dice on the first throw which was not 7 or 11, and not 2, 3, or 12. If $R = F$, then the computer has won the bet. The machine continues to execute the program by telling you that you have lost the bet (Step 490).

If the value of R is not 7 and is not F, then in Step 470 the computer returns to Step 450 and executes the subroutine a third time. The subroutine of Steps 710 to 750 is executed until one of the conditions of Steps 460 and 470 is satisfied.

e. After the computer prints the message stating that you have either won or lost the bet, it assigns the value of $T = T+B$ or $T = T-B$ in Steps 540 and 500 respectively.

Recall the ON-GO TO command of Section 5-5, and the SGN function of Section 4-1. If the present value of the cumulative winnings T is positive, then $SGN(T) = 1$. Hence, $SGN(T) + 2 = 3$. Therefore the command ON $SGN(T) +2$ GO TO 570, 590, 610 directs the computer to Step 610, and the computer prints the message YOUR TOTAL WINNINGS ARE and the present cumulative value of T.

Similarly, if $T<0$, then $SGN(T) = -1$ and $SGN(T)+2 = 1$. Hence, ON $SGN(T)+2$ GO TO 570, 590, 610 sends the computer to Step 570. Also, if $T = 0$, then the computer jumps to Step 590 from the ON-GO TO command.

f. After each game is played, the program jumps to Step 630, three lines are skipped in the OUTPUT, and the computer asks the question DO YOU WANT TO PLAY AGAIN ?. If you answer YES, or any string beginning with Y or Z (because of the relation $>=$ in Step 680), then the computer jumps back to Step 270 and the program is executed again, beginning at Step 270. Note that executing the program from Step 270 avoids repeating the directions for playing the game.

If you answer NO in response to the question of Step 660, or any string not beginning with Y or Z, the computer stops the execution of the program because of the STOP command of Step 690. The STOP command is necessary to prevent the computer from executing the subroutine of Steps 710 to 750 again.

5. Facts About the GOSUB-RETURN Commands

a. A GOSUB statement may appear within a subroutine. When this occurs, we say that we have *nested* subroutines. Let us designate the subroutine called by the first GOSUB in a program as Subroutine 1, and the subroutine called by the second GOSUB as Subroutine 2. During execution of the program, the first GOSUB encountered calls Subroutine 1, which contains another GOSUB command. This second GOSUB statement calls Subroutine 2. The computer executes Subroutine 2 until it encounters a RETURN statement. The machine then returns to the step following the GOSUB command in Subroutine 1. The next RETURN statement encountered sends the computer back to the main program. You shall see an example of nested GOSUB commands in Section 6-3.

b. It is important to realize that the computer will exit from a subroutine only by a RETURN statement and *not* by a GO TO, IF-THEN, or ON-GO TO statement.

c. There may be more than one RETURN statement in a subroutine. The first RETURN encountered sends the computer back to the step following the GOSUB statement which called that subroutine.

d. A GOSUB statement does not require a smaller step number than the RETURN statement. The GOSUB statement must be executed before the RETURN however.

EXERCISES 6-1

1. When are subroutines useful? Explain the distinction between the GOSUB and GO TO commands.
2. Write a program which asks for a number and then prints that number rounded off to 4 decimal places, to 1 decimal place, and to an integer. Use a subroutine to round off the number.
3. Using a subroutine which chooses randomly either 0 or 1, write a program which finds the average of 10, 30, and 100 such random selections. How do these averages compare?

6-2 The MAT INPUT, NUM Commands: Program (8); Sum, Product, Maximum, Minimum

The program given in this section introduces you to the statements MAT INPUT and NUM. The program also develops the routines for finding the sum and product of a set of numbers, and the maximum and minimum numbers in that set. These routines will prove useful to you when you are programming other problems which require one or more of these ideas.

1. MAT INPUT and NUM

The MAT INPUT command is similar to the INPUT statement, except that MAT INPUT permits us to enter any number of values, as you will see in Program (8).

The general form of the MAT INPUT statement is

MAT INPUT V ,

where V is a single-letter numeric or string variable. Variables such as A2 and C3$ are not allowed in the MAT INPUT statement.

The first value we type in response to the question mark of the MAT INPUT statement is assigned to the variable $V(1)$, the second value we type is assigned to V(2), the third to V(3), etc. After we have inputed all our data, NUM is set equal to the number of values entered.

The MAT INPUT statement allows us to enter data on more than one line. By typing an ampersand (&) at the end of each line of data before striking the RETURN key, we can enter as many values as we wish. We must, of course, reserve sufficient space in the DIM statement of the variable V.

If we do not type any values for the MAT INPUT statement, but simply hit the RETURN key after the question mark, then NUM is set equal to zero.

Step 150 of Program (8) utilizes the MAT INPUT command. The statement MAT INPUT A assigns the first value we enter to the variable A(1), the second value to A(2), ..., the Nth value to A(N). Notice at the end of the RUN of Program (8) that the input data was not limited to one line. At the end of the first line we typed an ampersand & before we struck the RETURN key. We then continued to input data on the next line. We can input as much data as we like, using this technique. Note that if a step contains the command INPUT (with no MAT), and we type & at the end of that data line, the computer will respond with an appropriate ERROR message.

In order to calculate the sum and product of a set of numbers and to find the maximum and minimum numbers in that set, it is necessary to know the number of elements in the set. We do not need to write a separate routine in Program (8) asking the user to specify the number of values he wishes to use, since the function NUM in Steps 240, 310, 380, and 460 automatically tallies this number for us.

PROGRAM (8):

```
10 DIM A(100)
20 PRINT "   THIS PROGRAM FINDS THE SUM, PRODUCT, MAXIMUM OR MINIMUM"
30 PRINT "OF AN ARRAY OF NUMBERS."
40 PRINT "   WHEN THE COMPUTER ASKS YOU, TYPE THE NUMBERS IN YOUR ARRAY"
50 PRINT "SEPARATED BY COMMAS.  IF YOU NEED MORE THAN ONE LINE FOR YOUR "
60 PRINT "DATA, TYPE '&' BEFORE YOU HIT THE RETURN KEY."
70 PRINT
80 REM----------SUM, PRODUCT, MAXIMUM OR MINIMUM?----------
90 PRINT "WHAT DO YOU WANT TO FIND ";
100 INPUT Q$
110 PRINT
120 PRINT
130 REM----------ENTERING NUMBERS----------
140 PRINT "WHAT ARE THE NUMBERS ";
150 MAT INPUT A
160 PRINT
```

6-2 The MAT INPUT, NUM Commands; Program (8)

```
170 PRINT
180 REM---------BRANCHING----------
190 IF Q$="PRODUCT" THEN 300
200 IF Q$="MAXIMUM" THEN 370
210 IF Q$="MINIMUM" THEN 450
220 REM---------SUM----------
230 S=0
240 FOR I=1 TO NUM
250    S=S+A(I)
260 NEXT I
270 PRINT "THE SUM IS "S
280 GOTO 520
290 REM----------PRODUCT---------
300 P=1
310 FOR I=1 TO NUM
320    P=P*A(I)
330 NEXT I
340 PRINT "THE PRODUCT IS "P
350 GOTO 520
360 REM---------MAXIMUM----------
370 H=A(1)
380 FOR I=2 TO NUM
390    IF H>A(I) THEN 410
400    H=A(I)
410 NEXT I
420 PRINT "THE MAXIMUM IS "H
430 GOTO 520
440 REM---------MINIMUM----------
450 L=A(1)
460 FOR I=2 TO NUM
470    IF L<A(I) THEN 490
480    L=A(I)
490 NEXT I
500 PRINT "THE MINIMUM IS "L
510 REM---------REPEAT?----------
520 PRINT
530 PRINT
540 PRINT "DO YOU WISH TO RUN AGAIN ";
550 INPUT R$
560 IF R$>="Y" THEN 70
570 END
```

OUTPUT:

   THIS PROGRAM FINDS THE SUM, PRODUCT, MAXIMUM OR MINIMUM
OF AN ARRAY OF NUMBERS.
   WHEN THE COMPUTER ASKS YOU, TYPE THE NUMBERS IN YOUR ARRAY
SEPARATED BY COMMAS. IF YOU NEED MORE THAN ONE LINE FOR YOUR
DATA, TYPE '&' BEFORE YOU HIT THE RETURN KEY.

WHAT DO YOU WANT TO FIND ?SUM

WHAT ARE THE NUMBERS ?2,4,6,8,10

THE SUM IS 30

DO YOU WISH TO RUN AGAIN ?YES

```
WHAT DO YOU WANT TO FIND ?PRODUCT

WHAT ARE THE NUMBERS ?1,2,3,4,5

THE PRODUCT IS   120

DO YOU WISH TO RUN AGAIN ?YES
WHAT DO YOU WANT TO FIND ?MAXIMUM

WHAT ARE THE NUMBERS ?23,-4632,37825,2.43,5.6E10

THE MAXIMUM IS   5.60000 E+10

DO YOU WISH TO RUN AGAIN ?YES
WHAT DO YOU WANT TO FIND ?MINIMUM

WHAT ARE THE NUMBERS ?46,384,-2837,5.74E-10,-5.74E10

THE MINIMUM IS  -5.74000 E+10

DO YOU WISH TO RUN AGAIN ?Z
WHAT DO YOU WANT TO FIND ?SUM

WHAT ARE THE NUMBERS ?23,415,10.21,-16.3,1.29E2,291.336,-601.2&
NOT ENOUGH INPUT--ADD MORE
?22.6,17.8,-34.445,1E-3

THE SUM IS   257.002

DO YOU WISH TO RUN AGAIN ?NO
```

---

2. Comments about Program (8)

a. Notice that Steps 190 to 210 direct the computer to other parts of the program, dependent on the string assigned to the string variable Q$ in Step 100. For example, if we type MINIMUM in response to Step 100, the conditional statements of Steps 190 and 200 are not satisfied. The condition of Step 210 is satisfied, however. Hence the computer jumps to Step 450.

Note that a statement such as IF Q$ = "SUM" THEN N is not

necessary in the branching section of the program. If we type SUM in response to the question of Step 90, or any string other than PRODUCT, MAXIMUM or MINIMUM, then none of the conditions of Steps 190 through 210 is satisfied. Thus, the computer continues to execute the steps in numerical order and finds the SUM of the input numbers by executing Steps 230 through 270.

b. Notice that Program (8) is written in parts, each of which performs a specific task. Thus, Steps 230 to 270 comprise the part which computes and types the SUM of the numbers, Steps 300 to 340 compute and type the PRODUCT, Steps 370 to 420 the MAXIMUM, and Steps 450 to 500 the MINIMUM. Steps 190 to 210 may be considered the control part of the program.

It is good programming practice to arrange your program in parts, or *routines,* with each part performing a specific task, and one part controlling all the others. Also, the use of REM statements and apostrophes help considerably in organizing and correcting *(debugging)* a program. More will be said about debugging a program in Section 7-1.

c. Notice that a GO TO 520 statement is included at the end of each part, in Steps 280, 350, and 430. The routine in Steps 520 through 560 asks if you wish to repeat the program. If you answer YES, or any string beginning with Y or Z (because of the relation >= in Step 560), then the computer jumps back to Step 70 and the program is executed again from Step 70. If you answer NO in response to the INPUT statement of Step 550, or any string not beginning with Y or Z, then the execution of the program stops, since the condition of Step 560 is not satisfied.

d. In Step 230, we could have set S equal to A(1) and changed Step 240 to FOR I = 2 TO NUM, thus saving one loop of the subroutine. We have not done this, however, so that you can easily adapt the SUM routine to your own programs. Similarly, we did not set P equal to A(1) in Step 300 and change Step 310 to FOR I = 2 TO NUM.

EXERCISES 6-2

1. Write a program which converts a positive "integral" number from any base B to base 10. Allow for repetition.
2. Write a program which converts a positive integer in base 10 to any base B. Allow for repetition.

3. Write a program which converts a positive "integral" number from any base B1 to any other base B2.

4. Write a program which will find the t_n term of an arithmetic sequence $t_n = a+(n-1)d$, for a given value of the first term a, the common difference d, and the number of the term n.

5. Write a program which will find the sum of an arithmetic series $a+(a+d)+(a+2d)+\ldots+(a+(n-1)d)$, for a given value of a, d, and n.

6. Write a program which will find the t_n term of a geometric sequence $t_n = a \cdot r^{n-1}$, for a given value of the first term a, the common ratio r, and the number of the term n.

7. Write a program which will find the sum of a geometric series $a+ar+ar^2+ar^3+\ldots+ar^{n-1}$, for a given value of a, r, and n.

8. Write a program which types the coefficient of the rth term of a binomial expansion $(a+b)^n$, where $1 \leqslant r \leqslant n$.

9. Write a program which finds the sum S_n of the series

$$\frac{1}{1 \cdot 2} + \frac{1}{2 \cdot 3} + \frac{1}{3 \cdot 4} + \cdots + \frac{1}{n(n+1)},$$

for the first n terms. What number is S_n approaching as n increases?

6-3 The CHANGE Command; Program (9): Algebra Quiz

1. The ASCII Code

Each character on the keyboard of the Teletype has a number associated with it. It is the equivalent of this number which comprises the electronic signal which enables you to interchange information with the computer. Many manufacturers of computer and input/output devices have agreed to use the same number code, called ASCII: American Standard Code for Information Interchange.

2. The CHANGE Command

Some time-sharing systems which use the BASIC language have taken advantage of the ASCII code numbering scheme by incorporating into their system a command called CHANGE. The CHANGE command converts each character of a string into its respective ASCII code number, or a code number into its corresponding character.

The general form of the CHANGE command is

CHANGE V$ TO V

or

CHANGE V TO V$,

where V$ is any string variable and V is an unsubscripted single-letter numerical variable. (Variables for V such as F2 and H(6) are not allowed.)

The main program in this section — Algebra Quiz — not only illustrates the CHANGE command, but also represents an attempt to explore that subject broadly referred to as Computer Assisted Instruction. However, a thorough understanding of the CHANGE command is first necessary before that program can be fully appreciated.

Look at Example (1) below. The CHANGE command in Step 30 converts each character of the string variable B$ into its corresponding ASCII code number. Thus, in the OUTPUT of Example (1), you see that the first character A has associated with it the number 65, B is changed to 66, C to 67, ..., Z to 90.

Step 10 DIMensions the subscripted variable A(I) for I = 0 to I = 30. That is, the computer reserves space for thirty-one subscripted variables A(0), A(1), A(2), ..., A(30), The command CHANGE B$ TO A in Step 30 causes the computer to change the first element in the string data, A, to its ASCII code number, 65, and store that number in the space A(1); then to change the second element B to its code number, 66, and store it in A(2), and so on, until the last element Z is changed to 90 and stored in A(26).

EXAMPLE (1)

PROGRAM:

```
10 DIM A(30)
20 READ B$
30 CHANGE B$ TO A
40 PRINT "THE NUMBER OF STRING CHARACTERS IS"A(0)
50 PRINT "THE CODE NUMBERS OF THE CHARACTERS IN THE"
55 PRINT "STRING "B$;" ARE:"
60 FOR I=1 TO A(0)
70    PRINT A(I);
80 NEXT I
90 PRINT
100 DATA ABCDEFGHIJKLMNOPQRSTUVWXYZ
110 END
```

OUTPUT:

```
THE NUMBER OF STRING CHARACTERS IS 26
THE CODE NUMBERS OF THE CHARACTERS IN THE
STRING ABCDEFGHIJKLMNOPQRSTUVWXYZ ARE:
 65  66  67  68  69  70  71  72  73  74  75  76  77  78  79  80  81
 82  83  84  85  86  87  88  89  90
```

The computer automatically counts the number of elements in the string (26 in Example (1)), and stores that number in space A(0). That is why you see the number 26 in the OUTPUT of Step 40 in Example (1). Steps 50 and 55 print a message, including the value of the string B$, and Steps 60 through 80 print the values of the subscripted variables A(1), A(2), ..., A(26).

The PRINT command in Step 90 directs the computer to return the carriage and line feed the paper. This PRINT statement is necessary because the semicolon in Step 70 suppresses the RETURN and LINE FEED directions each time the loop of Steps 60 through 80 is executed, thus printing the values of A(1), A(2), ..., A(26) horizontally. After the last value A(26) is printed, the computer executes Step 90. The carriage is then returned, and the paper is moved one line.

3. ASCII Code Numbers

Example (1) illustrated how we can change the letters of a string into their corresponding ASCII code numbers. Example (2) below illustrates how we can change a set of ASCII code numbers into their corresponding characters.

In converting code numbers to their corresponding characters, we assign the code number of the first character to a variable V(1), the code number of the second character to V(2), ..., the code number of the Nth character to V(N). We set V(0) equal to the number of code numbers we wish to convert. The command CHANGE V TO V$ assigns a string to the string variable V$. The string consists of V(0) characters whose code numbers are the values of V(1), V(2), ..., V(V(0)).

EXAMPLE (2)

PROGRAM:

```
10 PRINT "CHARACTER","CODE NUMBER"," ","CHARACTER","CODE NUMBER"
20 PRINT
30 C(0)=D(0)=1
40 FOR N=32 TO 63
50    C(1)=N
60    D(1)=N+32
70    CHANGE C TO F$
73    IF F$<>" " THEN 80
77    F$="(SPACE)"
80    CHANGE D TO G$
90    PRINT F$,C(1)," ",G$,D(1)
100 NEXT N
110 END
```

6-3 The CHANGE Command; Program (9)

OUTPUT:

CHARACTER	CODE NUMBER	CHARACTER	CODE NUMBER
(SPACE)	32	@	64
!	33	A	65
"	34	B	66
#	35	C	67
$	36	D	68
%	37	E	69
&	38	F	70
'	39	G	71
(	40	H	72
)	41	I	73
*	42	J	74
+	43	K	75
,	44	L	76
-	45	M	77
.	46	N	78
/	47	O	79
0	48	P	80
1	49	Q	81
2	50	R	82
3	51	S	83
4	52	T	84
5	53	U	85
6	54	V	86
7	55	W	87
8	56	X	88
9	57	Y	89
:	58	Z	90
;	59	[	91
<	60	\	92
=	61	]	93
>	62	↑	94
?	63	←	95

The ASCII code numbers range from 0 to 127. Example (2) generates a list of the most useful characters and their corresponding code numbers. In Step 30 of Example (2), we set the counters C(0) and D(0) to 1. That is, the number of characters in each string of the OUTPUT will be 1. In the first execution of the loop of Steps 40 to 100, C(1) = 32. The CHANGE command of Step 70 changes the ASCII code number 32 to its corresponding character (a space), and stores it as the value for the string variable F$. Then, since the condition of Step 73 is not satisfied, the computer continues to Step 77 and redefines F$ as "(SPACE)". Step 80 changes 64 to its corresponding character (@), and stores it as the value for G$. The computer continues to Step 90, prints the value of F$, the value of C(1), skips to the fourth column, prints the value of G$, the value of D(1), and then returns the carriage and line feeds the paper. The computer executes the loop a second time with N equal to 33. Steps 50 and 60 set C(1) equal to 33 and D(1) equal to 65. The CHANGE command changes 33 to its corresponding character (!) and stores it as the new value for F$. Similarly, 65 is changed to its corresponding ASCII character (the letter A), and it is stored as the new value for G$. The computer prints this new value of F$ and the value of C(1), skips a column, and then prints the new value of G$ and D(1). The computer executes this loop until N = 63, at which time the program is completed. Two string variables F$ and G$ were used instead of only one so that the table in the OUTPUT would be more compact.

4. Other Useful ASCII Code Numbers

Other ASCII code numbers which you might find useful and which are not included in the OUTPUT of Example (2) above are:

Character	Code Number
CONTROL D (EOT= End of Transmission)	4
CONTROL G (Bell)	7
LINE FEED	10
CARRIAGE RETURN	13
RUB-OUT (For Paper Tape)	127

The code numbers not included in Paragraphs 3 and 4 above correspond to: (1) lower-case letters on some Teletypes; (2) a repetition of some characters listed in Example (2) above on some Teletypes; (3) no characters on some Teletypes. You can easily generate a complete table of ASCII numbers and their corresponding characters by modifying the program of Example (2). Do not be surprised if the computer disconnects your Teletype when printing the character corresponding to the number 4!

5. Extracting Characters from a String

Just as the INT function can be used to isolate digits in a number, the CHANGE command is frequently used to extract characters from a string, as in Example (3) below. That is, the computer first asks you to input a string (Step 30). It then asks you to select a character in that string (Step 60). The computer then tells you the number of times the selected character appears in the string.

In the first part of the OUTPUT of Example (3), we entered the string MISSISSIPPI for the string variable A$, and the character I for B$. In Step 70 the characters of A$ are converted to their ASCII code numbers. Thus $S(1) = 77$ (the code number for the character M), $S(2) = 73$ (the code number for I), $S(3) = 83$ (the code number for S), ..., $S(11) = 73$. $S(0)$ is set equal to 11, since there are 11 characters in the string. Step 80 sets $C(0)$ equal to 1 (the number of characters in the string variable B$), and $C(1)$ equal to 73 (the ASCII code number for the letter I).

Now that we have converted all the characters of MISSISSIPPI to numbers, we can easily find the number of I's the string contains. The computer counts the number of variables from $S(1)$ to $S(11)$ which are equal to 73 using the loop of Steps 100 through 130, and then prints the sum in Step 140.

The characters in the other two strings of the OUTPUT are counted in a similar manner to that described above.

EXAMPLE (3)

PROGRAM:

```
10 DIM S(100)
20 PRINT "STRING ";
30 INPUT A$
40 IF A$="STOP" THEN 170
50 PRINT "CHARACTER ";
60 INPUT B$
70 CHANGE A$ TO S
80 CHANGE B$ TO C
90 N=0
100 FOR I=1 TO S(0)
110     IF S(I)<>C(1) THEN 130
120     N=N+1
130 NEXT I
140 PRINT "THERE ARE"N;B$;"'S IN YOUR STRING."
150 PRINT
160 GOTO 20
170 END
```

OUTPUT:

```
STRING ?MISSISSIPPI
CHARACTER ?I
THERE ARE 4 I'S IN YOUR STRING.

STRING ?SUBROUTINES HELP AVOID REPETITION OF STEPS IN A PROGRAM.
CHARACTER ?O
THERE ARE 5 O'S IN YOUR STRING.

STRING ?THE CODE NUMBER FOR A SPACE IS 32.
CHARACTER ?Z
THERE ARE 0 Z'S IN YOUR STRING.

STRING ?STOP
```

6. Program (9): Algebra Quiz

Program (9) is an algebra quiz administered, corrected, and graded by the computer. The program asks the student a question, corrects and comments on his answer, maintains a cumulative record of his responses, and tells him his percentile score at the end of the quiz, as you can see in the OUTPUT below. The CHANGE command is utilized so that the computer can decide whether or not the student's response to each question is correct.

The CHANGE command is used in the subroutine of Steps 560 through 730. The computer finds the sum of the ASCII code numbers of the correct answer as given in the DATA list, and the student's answer as typed on the keyboard. The computer then compares these two sums. If the student's answer is correct, then the sums will be equal, and the computer responds by typing an encouraging message. If the answer is not correct, then the sums will not be equal, and the computer types a message telling the student that he is wrong. The computer then types the correct answer. A detailed explanation of Program (9) is given in Paragraph 7 below.

PROGRAM (9):

```
10 DIM A(50)
20 RANDOM
30 PRINT TAB(15);"******ALGEBRA QUIZ******"
40 PRINT
50 PRINT
60 REM---------READING COMMENTS----------
70 FOR I=1 TO 10
80    READ C$(I)
90 NEXT I
100 FOR I=1 TO 5
110    READ D$(I)
120 NEXT I
130 REM---------QUOTATION MARK----------
140 F(0)=1
150 F(1)=34
160 CHANGE F TO G$
170 REM---------PART 1: PRODUCTS----------
180 PRINT "FIND THE PRODUCT OF EACH OF THE FOLLOWING."
190 PRINT "ENCLOSE YOUR ANSWER IN QUOTATION MARKS."
200 PRINT "FOR EXAMPLE: 3Y(2-X↑2)=";G$;"6Y-3(X↑2)Y"G$
210 GOSUB 410
220 REM---------PART 2: FACTORING----------
230 PRINT
240 PRINT
250 PRINT "FACTOR EACH OF THE FOLLOWING INTO PRIME FACTORS."
260 PRINT "FOR EXAMPLE: 16A-A(M↑4)=";G$;"A(M↑2+4)(2+M)(2-M)"G$
270 GOSUB 410
```

```
280 REM----------SCORE----------
290 PRINT TAB(12);"*****YOUR SCORE IS"10*N;"%*****"
300 IF N<8 THEN 380
310 ON N-7 GOTO 320,340,360
320 PRINT TAB(25);"GOOD!"
330 STOP
340 PRINT TAB(22);"VERY GOOD!"
350 STOP
360 PRINT TAB(22);"EXCELLENT!"
370 STOP
380 PRINT TAB(17);"YOU CAN DO BETTER."
390 STOP
400 REM----------GIVING TEST----------
410 PRINT
420 PRINT
450 FOR H=1 TO 5
460     READ Q$,A$,W$
470     PRINT Q$;
480     INPUT B$
490     GOSUB 550
500     PRINT
530 NEXT H
540 RETURN
550 REM----------CORRECTING QUIZ----------
560 IF B$=W$ THEN 710
570 CHANGE A$ TO A
580 S1=A(1)
590 FOR I=2 TO A(0)
600     S1=S1+A(I)
610 NEXT I
620 CHANGE B$ TO A
630 S2=A(1)
640 FOR I=2 TO A(0)
650     S2=S2+A(I)
660 NEXT I
670 IF S1<>S2 THEN 710
680 PRINT C$(INT(RND(Z)*10+1))
690 N=N+1
700 RETURN
710 PRINT D$(INT(RND(Z)*5+1))
720 PRINT "THE CORRECT ANSWER IS "A$
730 RETURN
740 REM-------COMMENTS, QUESTIONS, RIGHT AND WRONG ANSWERS-------
750 DATA YOUR ANSWER IS CORRECT!,GOOD!,RIGHT!,VERY GOOD!,WELL DONE!
760 DATA FINE,NICE GOING,CORRECT!,EXACTLY,EXCELLENT!
770 DATA YOUR ANSWER IS NOT CORRECT,NO!,SORRY,WRONG!,NOT QUITE
780 DATA "1. X(2X↑2-1)=","2X↑3-X","3X↑2-X"
790 DATA "2. (2X+1)(X-3)=","2X↑2-5X-3","3X↑2-5X-2"
800 DATA "3. (3X+Y↑2)(X-2Y↑2)=","-2Y↑4-5(Y↑2)X+3X↑2"
810 DATA                        "-2Y↑4+5(Y↑2)X-3X↑2"
820 DATA "4. (A↑X+B↑X)(A↑X-B↑X)=",A↑2X-B↑2X,"2A↑X-2B↑X"
830 DATA "5. (X↑N+2)↑2=",X↑2N+4X↑N+4,"2X↑N+4X↑N+4"
840 DATA "6. C↑2-81=","(C+9)(C-9)","(9+C)(9-C)"
850 DATA "7. 1-64M↑2=","(1+8M)(1-8M)","(8M+1)(8M-1)"
860 DATA "8. C↑3-3C↑2+3C-9=","(C↑2+3)(C-3)","(C↑2-3)(C+3)"
870 DATA "9. A(Y↑4)+2A(Y↑2)-24A=",A(Y↑2+6)(Y+2)(Y-2)
880 DATA                         A(Y↑2-6)(Y+2)(Y+2)
890 DATA "10. K↑3-27=","(K-3)(K↑2+3K+9)","(K+3)(K↑2-3K+9)"
900 END
```

OUTPUT:

```
******ALGEBRA QUIZ******


FIND THE PRODUCT OF EACH OF THE FOLLOWING.
ENCLOSE YOUR ANSWER IN QUOTATION MARKS.
FOR EXAMPLE: 3Y(2-X↑2)="6Y-3(X↑2)Y"


1. X(2X↑2-1)= ?"2X↑3-X"
NICE GOING

2. (2X+1)(X-3)= ?"2X↑2-5X-3"
YOUR ANSWER IS CORRECT!

3. (3X+Y↑2)(X-2Y↑2)= ?"-2Y↑4-7(Y↑2)X-3X↑2"
SORRY
THE CORRECT ANSWER IS -2Y↑4-5(Y↑2)X+3X↑2

4. (A↑X+B↑X)(A↑X-B↑X)= ?"A↑2X-B↑2X"
WELL DONE!

5. (X↑N+2)↑2= ?"X↑2N+4X↑N+4"
FINE



FACTOR EACH OF THE FOLLOWING INTO PRIME FACTORS.
FOR EXAMPLE: 16A-A(M↑4)="A(M↑2+4)(2+M)(2-M)"


6. C↑2-81= ?"(C-9)(C+9)"
WELL DONE!

7. 1-64M↑2= ?"(1-8M)(1+8M)"
RIGHT!

8. C↑3-3C↑2+3C-9= ?"(C↑2+3)(C-3)"
YOUR ANSWER IS CORRECT!

9. A(Y↑4)+2A(Y↑2)-24A= ?"A(Y↑2+6)(Y-2)(Y+2)"
FINE

10. K↑3-27= ?"(K+3)(K↑2-3K+9)"
YOUR ANSWER IS NOT CORRECT
THE CORRECT ANSWER IS (K-3)(K↑2+3K+9)

          *****YOUR SCORE IS 80 %*****
                     GOOD!
```

---

7. Comments about Program (9)

   a. Notice that Program (9) is separated into parts, with each part performing a specific task, and each part adequately labeled by means of a REM statement.

b. Steps 70 to 90 store the ten strings of encouraging remarks given in the DATA of Steps 750 and 760, and Steps 100 to 120 store the five comments given in the DATA of Step 770.

c. Look at the OUTPUT of Program (9). Notice the quotation marks typed at the end of the two sets of directions to the student. The statements contained in Steps 140 to 160 enable the computer to type these quotation marks. You have seen the CHANGE command of Step 160 similarly used in Example (2) of Paragraph 3 above. Step 140 of Program (9) specifies one element in a string; Step 150 assigns the ASCII code number for a quotation mark (34) to the variable F(1). Step 160 changes that code number into its corresponding character (") and stores it in G$. Steps 200 and 260 direct the computer to print the value of G$ in the appropriate place, and thus you see the quotation marks in the OUTPUT of the program.

d. After the computer types the directions for Part 1 of the quiz, it calls the subroutine (Step 210) beginning with Step 410. The loop contained in Steps 450 to 530: (1) reads the question (Q$), the correct answer (A$), and a wrong answer (W$) (Step 460 and the DATA of Step 780); (2) prints the question (Step 470); (3) asks for the student's answer (B$) (Step 480); and (4) calls the subroutine beginning with Step 550, which corrects the question (Step 490).

After the loop of Steps 450 to 530 has gone through the first five questions, the computer returns to Step 220, prints the directions to Part 2 of the quiz, and then calls the subroutine (Step 270) beginning in Step 410 for the last five questions.

e. The subroutine of Steps 560 to 730 corrects each question of the quiz. Step 560 determines whether or not the student's answer is the same as the wrong answer in the DATA. If the two strings are the same, the computer jumps to Step 710 and tells the student that he is wrong. If they are not the same, the computer continues to Step 570. The CHANGE command converts each character of A$ to its ASCII code number. For the first question of the quiz, A$ = 2X↑3-X. Look at the OUTPUT of Example (2) in Paragraph 3 above. The ASCII code number associated with the number 2 is 50, with X is 88, with ↑ is 94, with 3 is 51, with - is 45, and with X is 88. The CHANGE command of Step 570 assigns these values to A(I) for I = 1 to 6, as follows: A(1) = 50, A(2) = 88, A(3) = 94, A(4) = 51, A(5) = 45, A(6) = 88. Hence, after the sixth execution of the loop of Steps 590 to 610, S1 = 50+88+94+51+45+88 = 416.

Steps 620 to 660 find the sum S2 of the ASCII code numbers of the student's answer in a similar fashion.

Step 670 compares S1 and S2. If the value of S1 for Question 1 is not the same as the value of S2, then the computer jumps to Step 710, chooses a random number X between 1 and 5, prints the string contained in D$(X) – one of the five phrases which tells the student that he is wrong – and then prints the correct answer in Step 720. (On some systems the INT function in Step 710 is unnecessary, because subscripts are automatically truncated to integers.) The RETURN command in Step 730 sends the computer back to Step 500, and the process is repeated for Question 2.

If the value of S1 equals the value of S2 for Question 1, then the computer continues to Step 680, chooses a random number X between 1 and 10, prints the string contained in C$(X) – one of the ten phrases which tells the student that he answered the question correctly – increments by one the number N of correctly answered questions (Step 690), and then returns to Step 500 from the RETURN command of Step 700.

In Step 530, H is incremented by one, and the loop is repeated for Question 2.

After the first five questions are processed in this way, the computer continues the program by executing Step 540. The RETURN command sends the computer back to the step following the GOSUB command of Step 210.

The entire process of asking a question, checking the answer, and printing a message is repeated for Part 2 of the quiz. Notice that the program contains nested GOSUB commands in Steps 210 and 490 and in Steps 270 and 490. Also, in Steps 700 and 730 we used two RETURN statements in the two branches (Steps 680 to 700 and Steps 710 to 730) of the second of the nested subroutines (Steps 560 to 730).

f. Using the CHANGE command to correct the quiz allows the student considerable leeway in typing his answer. For example, the student may type (C-9)(C+9) or (C+9)(C-9) for the answer to Question 6, and his answer would be correct. If we had written the program directing the computer to compare the student's answer with the correct answer by using the equation A$ = B$, one of (C-9)(C+9) and (C+9)(C-9) would have been marked wrong. Thus, by using the CHANGE command, the only criterion for a correct answer is that the sum of the ASCII code numbers of the characters in the computer's answer and the student's answer be the same.

There is a potential pitfall to this method, however; one we must avoid. By using this comparison method, the computer interprets all answers with the same total ASCII code number as correct. For example, in Question 10, the computer would call the answer (K+3) (K↑2-3K+9) correct because the sum of the ASCII code numbers is the same as the sum of the code numbers in the correct answer (K-3) (K↑2+3K+9). Hence we must include a set of wrong responses in the DATA. That is why the third set of strings (for the string variable W$) is given in the DATA of Steps 780 to 890. Each string of this set is a wrong answer, but one whose sum of ASCII code numbers is the same as the correct answer. Thus in Step 560, before any comparison of ASCII code numbers is made in order to check the student's answer against the correct answer, the computer first compares the two strings contained in B$ and W$. If the strings are the same, then Step 560 sends the computer to Step 710, the beginning of the subroutine which processes a wrong answer.

g. Steps 290 to 390 calculate the student's score on the quiz, where N is the number of correct answers as determined in Step 690. Notice the ON-GO TO statement in Step 310: if N=8, then the computer jumps to Step 320 and prints GOOD!; if N=9, then the computer prints VERY GOOD! from Step 340; if N = 10, then the computer prints EXCELLENT! from Step 360.

EXERCISES 6-3

1. Write a program which asks for a string and then types that string backward.
2. Write a program which asks for a string and finds the number of blank spaces in that string.
3. Write a program which asks for a sentence and then types that sentence with no spaces between the words.
4. Write a program which asks for a sentence and then types that sentence with a carriage RETURN and LINE FEED after each word in the sentence.
5. Write a program which asks for a sentence and then types the number of words in that sentence.
6. Write a program which asks for a number and then types each digit of that number separately. Use the INT function.
7. Do Problem 6 using the CHANGE command.
8. Write a program which asks for several one-digit integers and then types these integers with no spaces between them. Use the CHANGE command.
9. Do Problem 8 without using the CHANGE command. (Hint: set A$(0) = "0", A$(1) = "1", A$(2) = "2", . . . , A$(9) = "9".)

130   Computer Programming in BASIC

10. Write a program which converts a number from base 10 to any base.

11. Write a program which determines which vowel is used most frequently in a string. The program should contain ten sentences in DATA statements.

12. Write a program which counts the number of times the word THE is used in ten sentences which are contained in DATA statements.

13. Write a program which asks for a word and counts the number of times that word appears in ten sentences contained in DATA statements.

14. Write a program which types ten sentences contained in DATA statements. The program should return the carriage and line feed the paper after the computer types approximately sixty-five characters. The program should RETURN and LINE FEED at the end of a word, not between letters of a word. The lengths of the ten sentences should vary.

15. Write a 10-question quiz using some or all of Program (9), but with your own questions in the DATA.

6-4 Defining Functions

Although BASIC has most of the more useful functions built right into the system, there are many times when one wants to work with a function not built in. This section discusses how you can define and work with other functions.

1. The Value of a Function

Recall from your work in algebra that if a function f is defined by an equation, say $f(x) = 3x^2 - 5x + 1$, then $f(c) = 3c^2 - 5c + 1$. If $c = 2$, we have:

$$f(2) = 3(2)^2 - 5(2) + 1$$
$$= 12 - 10 + 1$$
$$= 3 \ .$$

We say that we have evaluated the function f at $x = 2$.

If, on the other hand, the function f were defined by the equation $f(x) = 37.28 \, x^{2/3} - 15.12 \, x^{1/3} + 17.14$, then the problem of computing $f(x)$ at $x = 2$ "by hand" is much more tedious. However, the computer can easily evaluate this and other kinds of functions f for which f(x) exists.

2. The Command DEF

The method by which you can define your own function in BASIC is as follows:

$$\text{DEF FN}*(X) = (\text{Your function}),$$

where * can be replaced by any letter of the alphabet. Thus you can define a maximum of twenty-six different functions in one program.

In Step 10 of Example (1) below we have initially defined a function FNA(X) = X. During execution the program asks you in Step 50 if you have defined your function. If you type NO (or any string whose first letter precedes Y in the alphabet because of the $<=$ relation in Step 70), then the computer jumps to Step 200 and gives you directions as to how you should define your function. After Step 210, the execution stops.

You should then follow the directions of Steps 200 and 210 by typing 10 DEF FNA(X) = (YOUR FUNCTION), $<$ RETURN $>$. Of course, what you have thus done is to replace the function of Step 10 with another function. The first function in Example (1) is FNA(X) = 2*X↑4-2*X+3.

You continue to follow the directions of Steps 200 and 210 by typing RUN and hitting the RETURN key. The computer then begins the execution of the program a second time. In Step 50, when the computer asks if you have defined your function, you now type YES (or any string beginning with Y or Z). The computer continues to the next Step 80, skips a line in the OUTPUT, and asks you the question of Step 90. After you input your value for X in Step 100, the computer types the value of FNA(X).

When the computer encounters the function name FNA in Step 130, it searches the program for the definition of FNA(X), substitutes the value of X for X in 2*X↑4-2*X+3 (the expression defining the function), evaluates this expression (2*1↑4-2*1+3 = 3), returns to Step 130, and types this value (3).

In Steps 160 to 180, the computer asks if you wish to repeat the program. If you want to evaluate a different function from the one you have in Step 10, you can redefine the function by following the directions of Steps 200 and 210 again. Notice in the OUTPUT of Example (1) that we used two different functions. Thus BASIC offers you considerable flexibility and ease in defining your own function.

EXAMPLE (1)

PROGRAM:

```
10 DEF FNA(X)=X
20 PRINT "THIS PROGRAM COMPUTES THE VALUE OF FNA(X) AT ANY POINT"
30 PRINT "FOR ANY DEFINED FUNCTION."
40 PRINT
50 PRINT "HAVE YOU DEFINED YOUR FUNCTION";
60 INPUT Q$
70 IF Q$<="Y" THEN 200
80 PRINT
90 PRINT "AT WHAT POINT DO YOU WANT THE VALUE OF THE FUNCTION";
100 INPUT X
110 PRINT
120 PRINT
130 PRINT "THE VALUE OF FNA(X) AT "X;"IS "FNA(X)
140 PRINT
150 PRINT
160 PRINT "DO YOU WISH TO RUN AGAIN";
170 INPUT A$
180 IF A$>="Y" THEN 40
190 STOP
200 PRINT "TYPE '10 DEF FNA(X)=(YOUR FUNCTION)', THEN PRESS"
210 PRINT "<RETURN> KEY, TYPE 'RUN', <RETURN>."
220 END
```

OUTPUT:

```
THIS PROGRAM COMPUTES THE VALUE OF FNA(X) AT ANY POINT
FOR ANY DEFINED FUNCTION.

HAVE YOU DEFINED YOUR FUNCTION ?NO
TYPE '10 DEF FNA(X)=(YOUR FUNCTION)', THEN PRESS
<RETURN> KEY, TYPE 'RUN', <RETURN>.


10 DEF FNA(X)=2*X↑4-2*X+3
RUN



THIS PROGRAM COMPUTES THE VALUE OF FNA(X) AT ANY POINT
FOR ANY DEFINED FUNCTION.

HAVE YOU DEFINED YOUR FUNCTION ?YES

AT WHAT POINT DO YOU WANT THE VALUE OF THE FUNCTION ?1


THE VALUE OF FNA(X) AT  1 IS  3

DO YOU WISH TO RUN AGAIN ?YES

HAVE YOU DEFINED YOUR FUNCTION ?NO
TYPE '10 DEF FNA(X)=(YOUR FUNCTION)', THEN PRESS
<RETURN> KEY, TYPE 'RUN', <RETURN>.
```

```
10 DEF FNA(X)=2*LOG(X)-3*COS(X)+5*X↑7
RUN
```

```
THIS PROGRAM COMPUTES THE VALUE OF FNA(X) AT ANY POINT
FOR ANY DEFINED FUNCTION.

HAVE YOU DEFINED YOUR FUNCTION ?YES

AT WHAT POINT DO YOU WANT THE VALUE OF THE FUNCTION ?3

THE VALUE OF FNA(X) AT   3 IS    10940.2

DO YOU WISH TO RUN AGAIN ?NO
```

3. Characteristics of User-defined Functions

On most systems user-defined functions have the following characteristics.

(1) The argument(s) may be any numerical expression or variable when the function is used. (In the DEF statement, however, the argument(s) must be an unsubscripted variable.)

(2) Functions may have any number of arguments, as Steps 10 and 20 of Example (2) below illustrate. Also, a function need not have any argument, as you will see in Example (5) below.

(3) A DEF statement may contain other functions which are either user-defined or built-in, as you can see in Steps 10 and 20 of Example (2) below.

(4) The DEF statement may appear anywhere in a program, as illustrated in Example (3) below.

(5) The DEFinition must appear on one line. However, one can have a multiline function definition with another form of the DEF statement, as you will learn in Paragraph 4 below.

(6) The DEF statement may contain variables other than the function's arguments, as you can see in Step 40 of Example (3). All variables which are not arguments of the function have their current value. Thus these variables can be given new values before the function is used again.

EXAMPLE (2)

PROGRAM:

```
10 DEF FNA(X,Y)=2*X↑2*Y-Y↑2+SIN(X-Y)
20 DEF FND(A,B,C,D)=FNA(A,B)-FNA(C,D)
30 PRINT FND(1,-2,2,-1)
40 END
```

OUTPUT:

1.00000

EXAMPLE (3)

PROGRAM:

```
10 READ A,B,C,D,Y
20 PRINT FNZ(1.16274)
30 DATA 3.1,-2.6,1.32,-0.21,1.526
40 DEF FNZ(X)=SIN(X*Y)-LOG(A+B)*ATN(X*C)-D
50 END
```

OUTPUT:

1.87788

4. Multiline Function Definitions

Definitions of functions which require more than one line can also be handled in BASIC, as you can see in Steps 20 to 100 of Example (4) below. The multiline function definition begins with the statement

DEF FN*(X)

and ends with the statement

FNEND .

The computer recognizes a multiline definition by the absence of an equal sign in the DEF statement. The statement FNEND (an abbreviation for Function End) indicates the end of the multiline function definition.

Recall that a function may appear anywhere in the program; this holds true for multiline function definitions as well.

Suppose that the SGN function were not included in the BASIC language. Using a multiline function definition, we could easily define our own SGN function, as in Example (4) below.

Step 20 introduces the function FNS and indicates the argument X. Since the first value of X is -6.3, the computer executes Step 40, jumps to Step 90, and assigns the value -1 to FNS. Note that the function name FNS in Step 90 now behaves like a variable FNS; in the definition we use a LET statement to assign a value to FNS. The computer then goes to Step 100, the Function End statement, and returns the number -1 as the value of the function FNS (-6.3).

The computer next evaluates FNS(0) in Step 10. Again the computer goes to the DEF FNS (X) statement in Step 20. Since X = 0, the computer continues to Step 50, assigns 0 to the variable FNS, jumps to Step 100, and returns 0 as the value of the function FNS. Similarly, the number 1 is returned as the value of the function FNS(9.7).

As the OUTPUT of Example (4) illustrates, the function FNS is the same as the SGN function included in BASIC.

Notice the indentation of Steps 30 to 90 in Example (4) to set off the multiline function definition. This practice will help you read your programs more easily.

EXAMPLE (4)

PROGRAM:

```
10  PRINT FNS(-6.3),FNS(0),FNS(9.7)
20  DEF FNS(X)
30     IF X>0 THEN 70
40     IF X<0 THEN 90
50     LET FNS=0
60     GOTO 100
70     LET FNS=1
80     GOTO 100
90     LET FNS=-1
100 FNEND
110 END
```

OUTPUT:

-1   0   1

A multiline function definition can also be used with no argument as part of the DEF statement, as in Example (5) below. No argument is necessary for FNC in Step 50 because the variable K in the multiline function definition is defined in the READ statement of Step 20. In Steps 60 to 90 FNC is used as a variable to calculate the value of the function. Note that we again use a LET statement to assign a value to FNC in Steps 60 and 80. (Of course, we can use a LET statement to assign values to FNC only inside a multiline definition.)

EXAMPLE (5)

PROGRAM:

```
10 FOR D=1 TO 5
20    READ K
30    PRINT K;"FACTORIAL IS"FNC
40 NEXT D
50 DEF FNC
60    LET FNC=1
70    FOR I=2 TO K
80       LET FNC=I*FNC
90    NEXT I
100 FNEND
110 DATA 0,3,5,10,20
120 END
```

OUTPUT:

```
0 FACTORIAL IS 1
3 FACTORIAL IS 6
5 FACTORIAL IS 120
10 FACTORIAL IS 3628800
20 FACTORIAL IS 2.43290 E+18
```

Example (5) above is a program which computes N factorial. Compare this approach to computing N factorial with that given in Program (6) of Section 5-7.

In the first loop of Steps 10 to 40 of Example (5), K has the value 0 from the READ-DATA statements of Steps 20 and 110 respectively. The function name FNC of Step 30 sends the computer to Step 50 to evaluate FNC. In Step 60 FNC is set equal to 1. Since K = 0 (the first entry of the DATA statement is 0), the loop of Steps 70 to 90 is not executed at all. (Recall from Section 5-7 that in a FOR-NEXT loop, if the initial value is greater than the final value for positive increment numbers, the loop is not executed.) Hence the computer goes to Step 100 and returns to Step 30, with 1 as the value of FNC. The computer completes Step 30 by printing 0 FACTORIAL IS 1, as you can see in the OUTPUT.

The control variable D is incremented by 1, and the loop of Steps 10 to 40 is executed again. Step 20 now reads 3 for the value of K. The computer evaluates FNC of Step 30 by jumping to the multiline function definition a second time. FNC is set equal to 1 in Step 60. Since K = 3, the computer executes the loop of Steps 70 to 90 two times. On the first loop, FNC = 2*1 = 2 and on the second loop, FNC = 3*2=6. The computer then goes on to Step 100, the function definition is completed by FNEND, and the computer executes Step 30 by printing 3 FACTORIAL IS 6.

The computer executes the loop of Steps 10 to 40 three more times before the program is completed, thus evaluating FNC for K = 5, K = 10 and K = 20, as you can see in the OUTPUT.

A multiline function definition can be very useful in manipulating functions in a manner similar to subscripted variables, as in Example (6) below.

The program of Example (6) finds the maximum value of five functions at a particular value of A. The formulas for the five functions are: $FNS(X,1) = X\uparrow 2-1$ (Step 100); $FNS(X, 2) = X+10$ (Step 120); $FNS(X, 3) = 1/X$ (Step 140); $FNS(X, 4) = X\uparrow 2-X$ (Step 160); and $FNS(X, 5) = 4*X$ (Step 180).

The variable A is set equal to 3 in the READ-DATA statements of Steps 10 and 200. In Step 20 the computer substitutes 3 for A and then evaluates FNS(3, 1) by jumping to the multiline definition of Steps 80 to 190. Since Y, the second argument in the function definition of Step 80, equals 1, the ON-GO TO statement sends the computer to Step 100. Substituting 3 for the first argument X, the computer sets FNS = 3↑2-1=9-1=8, jumps to Step 190, returns 8 as the value of FNS(A, 1) and assigns this value to M.

The loop in Steps 30 to 60 finds the maximum value of the functions, as in the SUM, PRODUCT, MAXIMUM and MINIMUM program of Section 6-2. In the first execution of the loop, I = 2. Step 40 compares the values of M and FNS(3,2). The computer goes to the DEF FNS (X, Y) statement of Step 80, continues to the ON-GOTO statement of Step 90, and jumps to Step 120, since Y=2. The computer then sets FNS = X+10=3+10=13, jumps to Step 190, and returns 13 as the value of FNS(3, 2). In Step 40, since the condition of the IF-THEN statement is not satisfied (M = 8 and FNS(3, 2) = 13), the computer continues to Step 50, setting M equal to 13.

The control variable I is incremented by 1, and the computer executes the loop of Steps 30 to 60 a second time by reading the FNS statement in Step 40, jumping to the multiline function definition, returning .333333 as the value of FNS(3, 3), and comparing M(13) with FNS(3, 3). This procedure continues until the loop has been executed four times. The computer then prints the message THE MAXIMUM VALUE AT 3 IS 13, as you can see in the OUTPUT.

EXAMPLE (6)

PROGRAM:

```
10 READ A
20 M=FNS(A,1)
30 FOR I=2 TO 5
40    IF M>=FNS(A,I) THEN 60
50    M=FNS(A,I)
60 NEXT I
70 PRINT "THE MAXIMUM AT"A;"IS"M
80 DEF FNS(X,Y)
90    ON Y GOTO 100,120,140,160,180
100   FNS=X↑2-1
110   GOTO 190
120   FNS=X+10
130   GOTO 190
140   FNS=1/X
150   GOTO 190
160   FNS=X↑2-X
170   GOTO 190
180   FNS=4*X
190 FNEND
200 DATA 3
210 END
```

OUTPUT:

THE MAXIMUM AT 3 IS 13

Note that we cannot exit from a multiline function definition by a GO TO, IF-THEN, or ON-GO TO statement. We can exit only by the statement FNEND. Also note that a DEF statement may not be placed inside a multiline function definition.

5. Functions and Subroutines

A function is similar to a subroutine in that in each case the computer branches to a particular section of the program, executes that part, and then returns to where it left off. There are, however, several differences between the two.

(1) There is a numerical value associated with a function, whereas a subroutine has no such value.

(2) Since a function has a definite value, it can be used in expressions. For example, in the statement LET M = FND(5)*X↑2, the computer finds the value of FND(5), multiplies that value by the value of X↑2, and places the result in storage area M.

(3) A function usually has arguments.

Functions, like subroutines, are used to avoid repetition in a program. Functions should be used when you need to perform the same operation on different values a number of times, such as when the factorial of a number is required, or when you must find the sum, product, maximum, or minimum of several arrays of numbers.

EXERCISES 6-4

1. Write a program which defines FNB(X) = X↑2+X-6, then types the values of FNB(X) for integral values of X from -5 to 4.

2. Write a program which asks for two numbers N and P, and then rounds off N to P places. Use a function FNR(N,P) to round off N.

3. Suppose BASIC contained no absolute value function. Define a function FNA(X) whose value is ABS(X).

4. Write a program which asks for a number N and then uses a multiline definition to find the sum of the numbers from 1 to N.

5. Write a program which finds the minimum value of five functions at X = A. Use a multiline definition.

6. Write a program which finds the sum of the values of ten functions at X = A.

7. Write a program which finds the product of the values of ten functions at X = A.

6-5 Program (10): To Find the Slope of F(X) at X = C

Another example of a program which utilizes the DEF command is given in this section. However, some mathematical background is first given to help you fully appreciate the program.

1. The Slope of F(X) at X=C

You are probably familiar with the definition of slope of a line from your first-year algebra course.

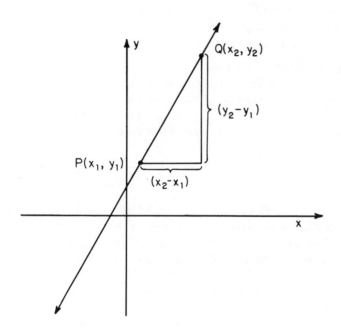

Figure (1)

In Figure (1) the slope of $\overleftrightarrow{PQ}$ is defined as follows:

$$M = \text{Slope}(\overleftrightarrow{PQ}) = \frac{y_2-y_1}{x_2-x_1}.$$

Let us extend the concept of slope to include functions whose graphs are not straight lines. For example, let us now consider the graph of the function f given by $f(x) = x^2$.

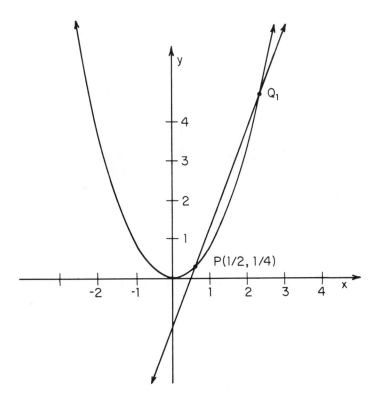

Figure (2)

Consider the point P (1/2, 1/4) and Q_1 any other point on the curve. The line $\overleftrightarrow{PQ_1}$ is called a *secant* line of the curve. If we let the point Q_1 approach the point P along the curve, we would have infinitely many secant lines $\overleftrightarrow{PQ_1}, \overleftrightarrow{PQ_2}, \ldots, \overleftrightarrow{PQ_n}, \ldots$, as shown in Figure (3).

As $Q_1, Q_2, Q_3, \ldots, Q_n, \ldots$ approach P along the curve, then the secant lines $\overleftrightarrow{PQ_1}, \overleftrightarrow{PQ_2}, \overleftrightarrow{PQ_3}, \ldots, \overleftrightarrow{PQ_n}, \ldots$ approach one line T, if it exists, called the *tangent* line.

The slope of the tangent line at $x = 1/2$ in Figure (3) is called the slope of the curve at $x = 1/2$. In general, the slope of a curve at a point $P(c, f(c))$ is the slope of the tangent line, if it exists, at $x = c$.

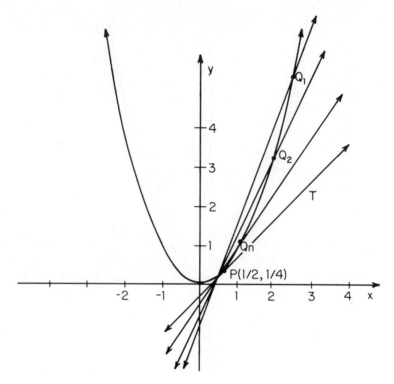

Figure (3)

2. Using the Computer to Find the Slope Number

How can we find the slope number of a function f at (c,f(c)) using the computer? We cannot find the slope number exactly, but we can find a rather close approximation to it. In fact, our approximation can be as accurate as we wish, up to the nine significant digit capability of the computer. (The number of significant digits varies with the system.)

Program (10), which computes the slope of the curve of a function F at X = C, follows the procedure discussed above: by taking two points P and Q on the curve, finding the slope of the secant line $\overleftrightarrow{PQ}$, and letting Q approach P, as in Figure (4).

In Figure (4), let a point P on the curve of a function F have coordinates (X,F(X)). Let J be a number such that (X+J) is the first coordinate of Q. The second coordinate of Q must then be F(X+J). In the right triangle PQR, R is a point such that the length of $\overline{RQ}$ is given by the number F(X+J) − F(X), and the length of $\overline{PR}$ is given by J.

6-5 Program (10): To Find the Slope of F(X) at X = C   143

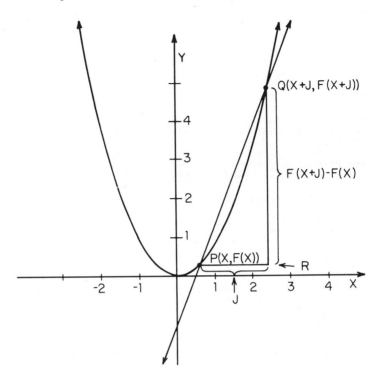

Figure (4)

The slope number S2 = $\dfrac{F(X+J)-F(X)}{J}$.

As Q approaches P along the curve, the value of J approaches 0. The slopes $S(1), S(2), \ldots, S(n), \ldots$ of the corresponding secant lines $\overleftrightarrow{PQ_1}, \overleftrightarrow{PQ_2}, \ldots, \overleftrightarrow{PQ_n}, \ldots$ approach the slope S of the tangent line T.

3. Program (10)

Program (10) begins on the next page.

4. Comments About Program (10)

a. Note the procedure by which you can define your own function, the directions for which are given in Steps 270 and 280 of Program (10). Recall that we used this procedure in Example (1) of Section 6-4; you should now be familiar with it.

PROGRAM (10):

```
10 DEF FNA(X)=X
20 PRINT "THIS PROGRAM FINDS THE SLOPE OF A FUNCTION AT ANY POINT."
30 PRINT "HAVE YOU DEFINED YOUR FUNCTION";
40 INPUT Q$
50 PRINT
60 IF Q$<"Y" THEN 270
70 REM----------ENTERING VALUE OF X----------
80 PRINT "AT WHAT POINT DO YOU WANT THE SLOPE";
90 INPUT X
100 PRINT
110 PRINT
115 REM----------FINDING SLOPE----------
120 LET J=1
130 LET S2=FNA(X+1)-FNA(X)
140 LET J=J/2
150 LET S1=S2
160 LET S2=(FNA(X+J)-FNA(X))/J
170 IF ABS(S1-S2)>=1E-5 THEN 140
180 PRINT "THE SLOPE OF FNA(X) AT "X;"IS "S2
190 PRINT
200 PRINT
210 REM----------REPEAT?----------
220 PRINT "DO YOU WISH TO RUN AGAIN";
230 INPUT A$
240 IF A$>="Y" THEN 30
250 STOP
260 REM----------INSTRUCTIONS FOR DEFINING FUNCTION----------
270 PRINT "TYPE '10 DEF FNA(X)=(YOUR FUNCTION)', THEN PRESS"
280 PRINT "<RETURN> KEY, TYPE 'RUN', <RETURN>."
290 PRINT
300 END
```

OUTPUT:

```
THIS PROGRAM FINDS THE SLOPE OF A FUNCTION AT ANY POINT.
HAVE YOU DEFINED YOUR FUNCTION ?NO

TYPE '10 DEF FNA(X)=(YOUR FUNCTION)', THEN PRESS
<RETURN> KEY, TYPE 'RUN', <RETURN>.



10 DEF FNA(X)=X↑2
RUN



THIS PROGRAM FINDS THE SLOPE OF A FUNCTION AT ANY POINT.
HAVE YOU DEFINED YOUR FUNCTION ?YES

AT WHAT POINT DO YOU WANT THE SLOPE ?3

THE SLOPE OF FNA(X) AT  3 IS  6

DO YOU WISH TO RUN AGAIN ?YES
HAVE YOU DEFINED YOUR FUNCTION ?NO
```

6-5 Program (10): To Find the Slope of F(X) at X = C

```
TYPE '10 DEF FNA(X)=(YOUR FUNCTION)', THEN PRESS
<RETURN> KEY, TYPE 'RUN', <RETURN>.


10 DEF FNA(X)=LOG(X)
RUN


THIS PROGRAM FINDS THE SLOPE OF A FUNCTION AT ANY POINT.
HAVE YOU DEFINED YOUR FUNCTION ?YES

AT WHAT POINT DO YOU WANT THE SLOPE ?2.7182818

THE SLOPE OF FNA(X) AT   2.71828 IS   0.367859

DO YOU WISH TO RUN AGAIN ?YES
HAVE YOU DEFINED YOUR FUNCTION ?NO

TYPE '10 DEF FNA(X)=(YOUR FUNCTION)', THEN PRESS
<RETURN> KEY, TYPE 'RUN', <RETURN>.


10 DEF FNA(X)=SIN(X)
RUN


THIS PROGRAM FINDS THE SLOPE OF A FUNCTION AT ANY POINT.
HAVE YOU DEFINED YOUR FUNCTION ?YES

AT WHAT POINT DO YOU WANT THE SLOPE ?1.57079

THE SLOPE OF FNA(X) AT   1.57079 IS   0

DO YOU WISH TO RUN AGAIN ?NO
```

---

b. After you have defined your function and have typed the value of X at which you want the slope determined (Steps 10 to 90), the computer continues to execute the program by finding the slope number at X in Steps 120 to 170. In Step 120, J is arbitrarily set to 1. Step 130 initializes S2 by setting it equal to the slope of the secant line for J = 1. Step 140 redefines J as 1/2. In Step 150, S1 is set equal to S2. Step 160 redefines S2 as the slope of the secant line for J = 1/2. Step 170 compares the values of S1 and S2.

If the two values differ by more than 1E-5, the computer jumps back to Step 140 and sets J equal to 1/4. (As you can see, the value of J is approaching 0.) S1 is set equal to S2 (the slope of the secant line for J = 1/2), and S2 is assigned the value of the slope of the secant line for

J = 1/4. The two values are again compared in Step 170. If their difference is still greater than 1E-5, the computer goes through the loop of Steps 140 to 170 a third time.

The computer continues to execute the loop until two consecutive slope numbers S1 and S2 differ by less than 1E-5. In each loop the value of J is divided by 2. As J approaches zero, the values of S1 and S2 (the slope numbers of two consecutive secant lines $\overset{\leftrightarrow}{PQ}$) approach the value of the slope of the tangent line T.

c. We prescribe the tolerance number 1E-5 in Step 170 to allow for the round-off error produced by the computer when it converts from decimal numbers to binary numbers. Thus in Program (10) we are assured that our answer is accurate to five decimal places. In Section 7-4 we shall show you a more sophisticated technique for allowing for round-off error.

EXERCISES 6-5

1. Write a program which DEFines a function FNL(X), and then find one ordered pair (X,FNL(X)) for which the graph of F has a maximum or a minimum value in some interval $A \leqslant X \leqslant B$.

2. Write a program which uses function definitions to define an Arcsin and an Arccos function from the built-in ATN function, and then types X, Arctan(X), Arcsin(X), and Arccos(X) for X = -1 to 1 in increments of 0.25.

3. Write a program which asks for a number X and a base B, and then types $LOG_B(X)$. Use the function FNL(X,B).

6-6 Summary

In this chapter you learned how to use the GOSUB-RETURN commands, the MAT INPUT statement and NUM, the CHANGE command, and the DEF command. Perhaps more importantly, you were exposed to a number of programs which gave you an opportunity to view these commands in different programming contexts. These programs also incorporated several techniques which should help you when you write your own programs.

You learned how to write a subroutine in a program with the use of the GOSUB-RETURN commands. You now know the distinction between a GO TO and a GOSUB command, and that you can have a set of nested GOSUB commands in a program.

You learned that MAT INPUT enables you to input more than one line of data with one command, and that NUM is equal to the number of entries you typed.

You learned that the CHANGE command converts each character on the keyboard to an ASCII code number, or a number to its corresponding character. This command allows us to manipulate characters in a string. The CHANGE command offered a method of writing a program in the area of Computer Assisted Instruction — an algebra quiz.

You now can define your own functions in a program by means of the DEF FN∗(X) command. You learned that you can also have multiline function definitions using the DEF FN∗(X) and FNEND commands. You became familiar with many of the characteristics of user-defined functions, both for single line and multiline function definitions.

Except for the MATrix statements which will be discussed in Chapter 8, there are no other commands on most BASIC systems with which you are not familiar. The material in Chapter 7 exposes you to programs in which you will see all of the commands you now know, but used in other contexts. Also, the Appendices consist of many programs dealing with topics from Algebra I through Calculus I, with additional miscellaneous topics.

CHAPTER 7

In this chapter you will first learn how to correct the mistakes you might make in a program, and then how to organize your program so that you can minimize the possibility of making an error. The remainder of the chapter is then concerned with three rather significant problems with which one deals in secondary school mathematics: (1) to find the remaining parts of a triangle, given any three parts (one of which must be a side); (2) to graph any well-defined function; and (3) to find the real zeros of any continuous function. Some mathematical background is first given for each of these problems, and then a program is presented by which the computer solves each problem.

7-1 Techniques of Debugging

1. Bugs

It is wishful thinking to expect that every program we write will RUN correctly the first time it is executed. There are usually some mistakes (bugs) which we must correct. The process of correcting mistakes in programs is called *debugging*.

There are two types of bugs which may appear in a program: (1) mistakes in the format of statements (usually errors in syntax or typographical mistakes); and (2) errors in the logical progression of statements.

2. ERROR Messages

The computer types ERROR messages during the compilation of a program and also during the execution of a program. Recall that during compilation the computer converts each statement in the program to the appropriate machine language instructions. Compilation occurs immediately after we type RUN but before execution begins. Compilation ERROR messages such as

ILLEGAL INSTRUCTION IN (Line Number)

inform us of syntactical errors in our program. We can correct such mistakes by correctly retyping the step indicated in the ERROR message.

For example, suppose a program for calculating the real roots of a quadratic equation was written as follows.

EXAMPLE (1)

PROGRAM:

```
10 PRINT "THIS PROGRAM CALCULATES THE REAL ROOTS OF A QUADRATIC ";
15 PRINT "EQUATION."
20 PRINT
30 PRINT "WHAT ARE THE COEFFICIENTS (START WITH THE X↑2 TERM AND"
40 PRINT "SEPARATE THE COEFFICIENTS WITH COMMAS)";
50 INPUR A,B,C
60 IF A<>0 THEN 90
70 PRINT "THE EQUATION IS LINEAR
80 GOTO 25
90 D=B↑2-4*A*C
100 IF D>=0 THEN 120
110 PRINT "THERE ARE NO REAL ROOTS IN YOUR EQUATION."
120 R1=(-B+SQR(D))/(2*A)
130 R2=(-B-SQR(D))/(2*A)
140 PRINT
150 PRINT "THE REAL ROOTS OF"A;"X↑2+"B;"X+"C;"ARE: "R1;","R2
160 PRINT "DO YOU WISH TO RUN AGAIN";
170 INPUT Q$
180 IF Q$>="Y" THEN 20
190 END
```

OUTPUT:

```
ILLEGAL INSTRUCTION IN 50
ILLEGAL FORMAT IN 70
UNDEFINED LINE NUMBER 25 IN 80
```

We see from the ERROR messages and the corresponding line numbers of Example (1) above that: (1) the word INPUT is misspelled in Step 50; (2) a quotation mark is missing in Step 70; and (3) we should have typed 20 instead of 25 in Step 80. Note that the program was not executed by the computer because of these errors.

We make the necessary changes in these three steps by retyping them while ON LINE, and then RUNning the program a second time.

CORRECTIONS IN PROGRAM OF EXAMPLE (1)

```
50  INPUT A,B,C
70  PRINT "THE EQUATION IS LINEAR."
80  GOTO 20
```

OUTPUT:

```
THIS PROGRAM CALCULATES THE REAL ROOTS OF A QUADRATIC EQUATION.

WHAT ARE THE COEFFICIENTS (START WITH THE X↑2 TERM AND
SEPARATE THE COEFFICIENTS WITH COMMAS) ?1,2,1

THE REAL ROOTS OF 1 X↑2+ 2 X+ 1 ARE: -1 ,-1
DO YOU WISH TO RUN AGAIN ?YES

WHAT ARE THE COEFFICIENTS (START WITH THE X↑2 TERM AND
SEPARATE THE COEFFICIENTS WITH COMMAS) ?1,2,6
THERE ARE NO REAL ROOTS IN YOUR EQUATION.
SQRT OF NEGATIVE NUMBER IN 120
SQRT OF NEGATIVE NUMBER IN 130

THE REAL ROOTS OF 1 X↑2+ 2 X+ 6 ARE:  1.23607 ,-3.23607
DO YOU WISH TO RUN AGAIN ?NO
```

In the above OUTPUT of the corrected version of Example (1) you see the second type of ERROR message: one which occurs during the execution of a program. The two execution ERRORs given in the OUTPUT above tell us that we are trying to take the square root of a negative number in Steps 120 and 130 of Example (1).

We thought we accounted for the possibility of a negative square root in Step 100. However, if $D < 0$, the computer will type the message of Step 110: THERE ARE NO REAL ROOTS IN YOUR EQUATION., but will then continue to execute the program by going on to Step 120, which requires the calculation of SQR(D). Since $D < 0$, the computer types the ERROR message, calculates SQR(ABS(D)) in finding R1, goes on to Step 130, types the same ERROR message for that step, uses SQR(ABS(D)) to calculate R2, and then continues to execute the program through to the END statement of Step 190.

To avoid the ERROR messages in the OUTPUT of Example (1), and to obtain the correct answer, we insert the following step.

CORRECTIONS IN PROGRAM OF EXAMPLE (1)

```
115 GOTO 160
```

By inserting Step 115 into the program of Example (1) while ON LINE, we have accounted for the case for which $D < 0$, in that the computer will now jump to Step 160, and you will be asked whether or not you wish to repeat the execution of the program.

We may obtain a complete LIST of the corrected version of Example (1) by typing the command LISTNH, as follows.

CORRECTED VERSION OF EXAMPLE (1)

PROGRAM:

```
LISTNH
10 PRINT "THIS PROGRAM CALCULATES THE REAL ROOTS OF A QUADRATIC ";
15 PRINT "EQUATION."
20 PRINT
30 PRINT "WHAT ARE THE COEFFICIENTS (START WITH THE X↑2 TERM AND"
40 PRINT "SEPARATE THE COEFFICIENTS WITH COMMAS)";
50 INPUT A,B,C
60 IF A<>0 THEN 90
70 PRINT "THE EQUATION IS LINEAR."
80 GOTO 20
90 D=B↑2-4*A*C
100 IF D>=0 THEN 120
110 PRINT "THERE ARE NO REAL ROOTS IN YOUR EQUATION."
115 GOTO 160
120 R1=(-B+SQR(D))/(2*A)
130 R2=(-B-SQR(D))/(2*A)
140 PRINT
150 PRINT "THE REAL ROOTS OF"A;"X↑2+"B;"X+"C;"ARE: "R1;",""R2
160 PRINT "DO YOU WISH TO RUN AGAIN";
170 INPUT Q$
180 IF Q$>="Y" THEN 20
190 END
```

OUTPUT:

```
THIS PROGRAM CALCULATES THE REAL ROOTS OF A QUADRATIC EQUATION.

WHAT ARE THE COEFFICIENTS (START WITH THE X↑2 TERM AND
SEPARATE THE COEFFICIENTS WITH COMMAS) ?1,5,4

THE REAL ROOTS OF 1 X↑2+ 5 X+ 4 ARE: -1 ,-4
DO YOU WISH TO RUN AGAIN ?YES

WHAT ARE THE COEFFICIENTS (START WITH THE X↑2 TERM AND
SEPARATE THE COEFFICIENTS WITH COMMAS) ?1,2,6
THERE ARE NO REAL ROOTS IN YOUR EQUATION.
DO YOU WISH TO RUN AGAIN ?NO
```

As you can see in the above OUTPUT, we have successfully debugged and executed the program of Example (1).

3. Errors in the Logical Progression of Statements

Mistakes which are included in a program but which the computer accepts as legal statements are much more difficult to find. Under these conditions, the program seems to RUN perfectly well, the computer gives us no ERROR messages, but we know that the answer is wrong!

For example, suppose we have written a program which finds the product of five numbers, as given in Example (2).

EXAMPLE (2)

PROGRAM:

```
10 A(0)=1
20 FOR I=1 TO 5
30     READ A(I)
40     P(I)=A(I-1)*A(I)
50 NEXT I
60 PRINT "THE PRODUCT IS"P(I)
70 DATA 3,5,7,9,11
80 END
```

OUTPUT:

```
THE PRODUCT IS 99
```

The program in Example (2) above ran properly, but we know that 3x5x7x9x11 is a number much larger than 99. In fact, computing the product "by hand" yields 10395.

In order to help us find the bug in the program of Example (2), we insert a PRINT statement inside the loop of Steps 20 through 50, as follows.

CORRECTION IN PROGRAM OF EXAMPLE (2)

```
45 PRINT I,A(I),P(I)
```

OUTPUT:

```
1             3              3
2             5             15
3             7             35
4             9             63
5            11             99
THE PRODUCT IS 99
```

In studying the values of the variables in the above OUTPUT of Example (2), we realize that the value of P(5) is the product of A(4) and A(5): 9 x 11. Similarly, P(4) is the product of A(3) and A(4): 7 x 9, etc. Our program in Example (2) is written in such a way that we lose all the previous products except the last two. What we thought the statement of Step 40 was calculating — the product of all five numbers — is incorrect.

We correct the program by rewriting Steps 10 and 40 and deleting Step 45, as follows.

CORRECTIONS IN PROGRAM OF EXAMPLE (2)

```
45
10 P(0)=1
40    P(I)=P(I-1)*A(I)
```

Note that we deleted Step 45 by simply typing the line number followed by ®.

Since we have made several changes in Example (2), we would like to see the program as it is presently written. We ask the computer to LIST the program by the command LISTNH.

CORRECTED VERSION OF EXAMPLE (2)

PROGRAM:

```
LISTNH
10 P(0)=1
20 FOR I=1 TO 5
30    READ A(I)
40    P(I)=P(I-1)*A(I)
50 NEXT I
60 PRINT "THE PRODUCT IS" P(I)
70 DATA 3,5,7,9,11
80 END
```

OUTPUT:

```
THE PRODUCT IS 10395
```

As you can see in the above OUTPUT, we have successfully debugged and executed the program of Example (2).

4. General Techniques of Debugging

Careful preparation of your program before you type your first OFF LINE draft will greatly reduce the number of mistakes you might make. The following general techniques will prove helpful when you write your programs.

   (1) Construct a flow chart of your program.

   (2) Write your program in parts or subroutines, as in our SUM, PRODUCT, MAXIMUM, and MINIMUM program.

   (3) Test each part "by hand," as if you were the computer executing the program.

   (4) Account for all possible cases in your program.

   (5) Use REM statements and apostrophes to remind you of the purpose of each part or subroutine.

7-2 Program (11): Solutions of Triangles in Trigonometry

If you have studied trigonometry, you will recall that given any three of the angles or sides (parts) of a triangle, one of which must be a side, you can compute the unknown angles or sides of the triangle by the Law of Sines or the Law of Cosines. There are a number of cases which you must consider, and the method by which you "solved" the triangle (calculated the remaining parts) was probably by logarithms. Let us briefly review each of these cases, and then we shall write one program which will handle all possibilities.

1. Case I: Angle-Side-Angle

Let us agree to label our triangle as follows.

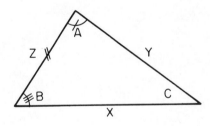

Figure (6)

The lengths X, Y, Z of the sides are measured in the same linear measure, and the measure of angles A, B, C is in degrees.

Suppose we were given A, Z, B as indicated in Figure (6). Since C = 180 − (A+B), by the Law of Sines we have:

$$\frac{X}{\text{Sin (A)}} = \frac{Z}{\text{Sin (C)}}.$$

Therefore,
$$X = \frac{Z \cdot \text{Sin (A)}}{\text{Sin (C)}}.$$

Similarly,
$$Y = \frac{X \cdot \text{Sin (B)}}{\text{Sin (A)}}.$$

We have thus found the remaining parts C, X and Y in terms of A, Z and B.

2. Case II: **Side-Angle-Side**

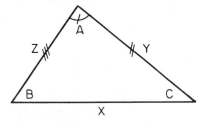

Figure (7)

Suppose we were given Y, A, Z as indicated in Figure (7). By the Law of Cosines, we have:

$$X^2 = Y^2 + Z^2 - 2YZ \cdot \text{Cos (A)}.$$

Therefore,
$$X = \sqrt{Y^2 + Z^2 - 2YZ \cdot \text{Cos (A)}}.$$

Now, by the Law of Sines, we have:

$$\frac{\text{Sin (B)}}{Y} = \frac{\text{Sin (A)}}{X}.$$

Hence,

$$\text{Sin}(B) = \frac{Y \cdot \text{Sin}(A)}{X}.$$

Therefore,

$$B = \text{Arcsin}\left(\frac{Y \cdot \text{Sin}(A)}{X}\right) \quad (1)$$

or

$$B = 180 - \text{Arcsin}\left(\frac{Y \cdot \text{Sin}(A)}{X}\right). \quad (2)$$

Since $C = 180 - (A + B)$, we have solved the triangle ABC.

However, since BASIC does not contain an Arcsin function, we must use a formula which expresses the Arcsin in terms of the Arctan function.

Recall the following figure from your work in trigonometry.

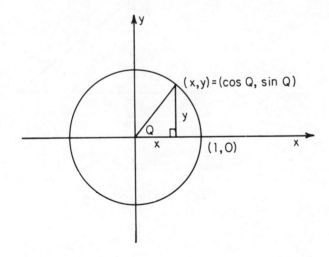

Figure (8)

From the definition of Arcsin, we have in Figure (8):

$$\text{Sin } Q = y \text{ such that Arcsin}(y) = Q$$

and

7-2 Program (11): Solutions of Triangles in Trigonometry

$$\text{Tan } Q = \frac{y}{x} \text{ such that } \text{Arctan}\left(\frac{y}{x}\right) = Q.$$

Therefore,

$$\text{Arcsin }(y) = \text{Arctan}\left(\frac{y}{x}\right).$$

Since $x = \sqrt{1-y^2}$ by the Pythagorean Theorem, we have:

$$\text{Arcsin }(y) = \text{Arctan}\left(\frac{y}{\sqrt{1-y^2}}\right). \qquad (3)$$

In Equation (1) above, let us define

$$D = \frac{Y \cdot \text{Sin }(A)}{X}.$$

Hence, another name for B in Equations (1) and (2) is:

$$B = \text{Arcsin }(D)$$
$$= \text{Arctan}\left(\frac{D}{\sqrt{1-D^2}}\right) \qquad (4)$$

or

$$B = 180 - \text{Arctan}\left(\frac{D}{\sqrt{1-D^2}}\right). \qquad (5)$$

3. Case III: Side-Side-Side

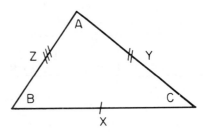

Figure (9)

Suppose we were given X, Y, Z as indicated in Figure (9). By the Law of Cosines we have:

$$X^2 = Y^2 + Z^2 - 2YZ \cdot \cos(A).$$

Since

$$\cos(A) = \frac{Y^2 + Z^2 - X^2}{2YZ},$$

then

$$A = \text{Arccos}\left(\frac{Y^2 + Z^2 - X^2}{2YZ}\right). \tag{6}$$

Similarly,

$$Y^2 = X^2 + Z^2 - 2XZ \cdot \cos(B).$$

Since

$$\cos B = \frac{X^2 + Z^2 - Y^2}{2XZ},$$

then

$$B = \text{Arccos}\left(\frac{X^2 + Z^2 - Y^2}{2XZ}\right). \tag{7}$$

Since $C = 180 - (A+B)$, we have calculated the three angles of triangle ABC.

But angles A and B are written in terms of the Arccos function in Equations (6) and (7) above. Since the Arccos function is not included in the BASIC language, we must convert Equations (6) and (7) into equations which include the Arctan function.

Look at Figure (8). Recall that

$$\text{Arccos}(y) = 90 - \text{Arcsin}(y).$$

But

$$\text{Arcsin}(y) = \text{Arctan}\left(\frac{y}{\sqrt{1-y^2}}\right)$$

from Equation (3) above. Therefore,

$$\text{Arccos}(y) = 90 - \text{Arctan}\left(\frac{y}{\sqrt{1-y^2}}\right). \tag{8}$$

In equation (6) and (7), let us respectively define

$$D_1 = \frac{Y^2 + Z^2 - X^2}{2YZ} \text{ and } D_2 = \frac{X^2 + Z^2 - Y^2}{2XZ}.$$

Hence, another name for A in Equation (6) is

$$A = 90 - \text{Arctan}\left(\frac{D_1}{\sqrt{1-D_1^2}}\right) \tag{9}$$

and another name for B in Equation (7) is

$$B = 90 - \text{Arctan}\left(\frac{D_2}{\sqrt{1-D_2^2}}\right). \tag{10}$$

4. Case IV: Side-Side-Angle (Ambiguous Case)

We are given the lengths of sides X and Y and the measure of angle A. The following possibilities exist.

a. Suppose that the measure of angle A is less than 90 degrees. Let H be the length of the perpendicular from the vertex opposite the side of length Z to that side, as shown in Figure (10) below.

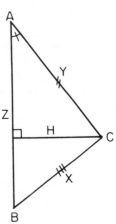

Figure (10)

Then, $\text{Sin}(A) = \frac{H}{Y}$ and $H = Y \cdot \text{Sin}(A)$. Under these conditions one of the following cases may occur.

(1) No triangle exists if $X < H$.
(2) One triangle exists if $X = H$, as illustrated in Figure (11).

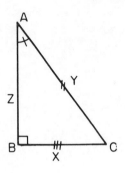

Figure (11)

In the right triangle ABC of Figure (11), we have:

$$X = H = Y \cdot \text{Sin}(A).$$

Since $m°(B) = 90$, then $Z = Y \cdot \text{Cos}(A)$ and

$$C = 180 - (B+A)$$
$$= 180 - (90+A)$$
$$= 90 - A.$$

Hence we have solved this particular case of triangle ABC.

(3) Two triangles ABC and A'B'C' exist if $H < X < Y$, as shown in Figure (12).

From the Law of Sines we have:

$$\frac{\text{Sin}(A)}{X} = \frac{\text{Sin}(B)}{Y}.$$

Therefore,

$$\text{Sin}(B) = \frac{Y}{X} \cdot \text{Sin}(A).$$

7-2 Program (11): Solutions of Triangles in Trigonometry   161

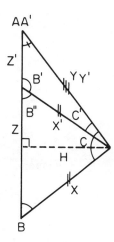

Figure (12)

Hence,

$$B = \operatorname{Arcsin}\left(\frac{Y}{X} \cdot \operatorname{Sin}(A)\right).$$

Since we now know the measures of angles B and A, we can easily find the measure of C by C = 180 − (A+B).

Using the Law of Sines again, we have:

$$\frac{X}{\operatorname{Sin}(A)} = \frac{Z}{\operatorname{Sin}(C)}.$$

Thus,

$$Z = \frac{X \cdot \operatorname{Sin}(C)}{\operatorname{Sin}(A)}.$$

We have found all parts of the triangle ABC. We must now solve the triangle A′ B′ C′.

Since X = X′ and B = B″, then

$$B' = 180 - B''$$

$$= 180 - B.$$

Also, since A = A′, then

$$C' = 180 - (A' + B')$$
$$= 180 - A - B'.$$

By the Law of Sines we have:

$$\frac{X'}{\text{Sin }(A')} = \frac{X}{\text{Sin }(A)} = \frac{Z'}{\text{Sin }(C')}.$$

Therefore,

$$Z' = \frac{X \cdot \text{Sin }(C')}{\text{Sin }(A)}.$$

We have thus found all parts of the second triangle A' B' C'.
(4) One triangle exists if $X > Y$, as shown in Figure (13).

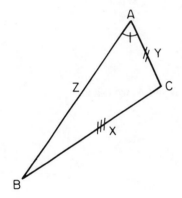

Figure (13)

Since

$$\frac{\text{Sin }(A)}{X} = \frac{\text{Sin }(B)}{Y}$$

then

$$B = \text{Arcsin}\left(\frac{Y}{X} \cdot \text{Sin }(A)\right).$$

Since we now know the measures of angles A and B, we can find C by $C = 180 - (A+B)$.

7-2 Program (11): Solutions of Triangles in Trigonometry

Now, since

$$\frac{X}{\text{Sin (A)}} = \frac{Z}{\text{Sin (C)}},$$

then

$$Z = \frac{X \cdot \text{Sin (C)}}{\text{Sin (A)}}.$$

We have thus found all of the remaining parts of triangle ABC for this case.

b. Suppose that the measure of angle A were greater than or equal to 90 degrees. The following cases may occur.

(1) One triangle ABC exists if $X > Y$, as shown in Figure (14).

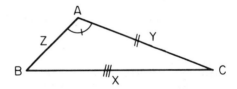

Figure (14)

By the Law of Sines we have:

$$\frac{\text{Sin (A)}}{X} = \frac{\text{Sin (B)}}{Y}.$$

Therefore,

$$B = \text{Arcsin}\left(\frac{Y}{X} \cdot \text{Sin (A)}\right).$$

Since

$$C = 180 - (A+B) \text{ and } \frac{X}{\text{Sin (A)}} = \frac{Z}{\text{Sin (C)}},$$

then

$$Z = \frac{X \cdot \text{Sin (C)}}{\text{Sin (A)}}.$$

We have thus solved triangle ABC in this case.
(2) No triangle exists if $X \leqslant Y$.

5. Traditional Solution of Triangle Problem

The solution of any one of the cases given in Paragraphs 1 through 4 above is a very difficult task without some tool for calculation, such as a desk calculator, or more commonly a table of logarithms. By either of these methods the calculations are quite tedious to perform, particularly if we require the lengths of sides to the nearest thousandth of a unit, say, and the angles to the nearest hundredth of a degree. However, a computer can handle such a problem with ease, as the following program testifies.

6. Program (11)

The program for the solution of the triangle problem is as follows.

```
1000 REM---------DIAGRAM----------
1010 PRINT "THIS PROGRAM FINDS THE REMAINING PARTS OF ANY TRIANGLE,"
1020 PRINT "GIVEN A SIDE AND ANY TWO OTHER PARTS."
1030 PRINT TAB(14);"."
1040 G=16
1050 FOR I=13 TO 3 STEP -1
1060    IF I=12 THEN 1120
1070    IF I=7 THEN 1140
1080    PRINT TAB(I);".";TAB(G);"."
1090    G=G+2
1100 NEXT I
1110 GOTO 1160
1120 PRINT TAB(12);". A";TAB(18);"."
1130 GOTO 1090
1140 PRINT TAB(5);"Z .";TAB(28);". Y"
1150 GOTO 1090
1160 PRINT TAB(2);". B";TAB(34);"C";TAB(38);"."
1170 PRINT " . . . . . . . . . . . . . . . ."
1180 PRINT TAB(18);"X"
1190 PRINT
1200 PRINT
1210 P1=3.14159/180
1220 REM---------WHICH CASE?----------
1230 PRINT "ARE YOU GIVEN ASA, SAS, SSS, OR SSA";
1240 INPUT Q$
1250 PRINT
1260 PRINT
1270 IF Q$="SAS" THEN 1430
1280 IF Q$="SSS" THEN 1580
1290 IF Q$="SSA" THEN 1770
1300 REM---------ASA----------
```

7-2 Program (11): Solutions of Triangles in Trigonometry

```
1310 PRINT "WHAT IS THE MEASURE OF ANGLE A IN DEGREES";
1320 INPUT A
1330 PRINT "WHAT IS THE MEASURE OF ANGLE B IN DEGREES";
1340 INPUT B
1350 PRINT "WHAT IS THE LENGTH OF SIDE Z";
1360 INPUT Z
1370 C=180-A-B
1380 A1=A*P1
1390 X=Z*SIN(A1)/SIN(C*P1)
1400 Y=X*SIN(B*P1)/SIN(A1)
1410 GOTO 2180
1420 REM----------SAS----------
1430 PRINT "WHAT IS THE LENGTH OF SIDE Y";
1440 INPUT Y
1450 PRINT "WHAT IS THE LENGTH OF SIDE Z";
1460 INPUT Z
1470 PRINT "WHAT IS THE MEASURE OF ANGLE A IN DEGREES";
1480 INPUT A
1490 A1=A*P1
1500 X=SQR(Y↑2+Z↑2-2*Y*Z*COS(A1))
1510 D=Y*SIN(A1)/X
1520 B=FNS(D)
1530 IF Z>=Y*SIN(A1) THEN 1550
1540 B=180-B
1550 C=180-A-B
1560 GOTO 2180
1570 REM----------SSS----------
1580 PRINT "WHAT IS THE LENGTH OF SIDE X";
1590 INPUT X
1600 PRINT "WHAT IS THE LENGTH OF SIDE Y";
1610 INPUT Y
1620 PRINT "WHAT IS THE LENGTH OF SIDE Z";
1630 INPUT Z
1640 D=(Y↑2+Z↑2-X↑2)/(2*Y*Z)
1650 IF ABS(D)<1 THEN 1700
1660 PRINT
1670 PRINT
1680 PRINT "NO TRIANGLE SATISFIES YOUR VALUES OF X, Y, AND Z."
1690 GOTO 2270
1700 A=FNC(D)
1710 D=(X↑2+Z↑2-Y↑2)/(2*X*Z)
1720 IF ABS(D)>=1 THEN 1660
1730 B=FNC(D)
1740 C=180-A-B
1750 GOTO 2180
1760 REM----------SSA----------
1770 PRINT "WHAT IS THE LENGTH OF SIDE X";
1780 INPUT X
1790 PRINT "WHAT IS THE LENGTH OF SIDE Y";
1800 INPUT Y
1810 PRINT "WHAT IS THE MEASURE OF ANGLE A IN DEGREES";
1820 INPUT A
1830 A1=A*P1
1840 IF A<90 THEN 1940
1850 REM---A>=80---
1860 IF X>Y THEN 1910
1870 PRINT
1880 PRINT
1890 PRINT "NO TRIANGLE SATISFIES YOUR VALUES OF X, Y, AND A."
1900 GOTO 2270
1910 GOSUB 2340
1920 GOTO 2180
1930 REM---A<90---
```

```
1940 H=SIN(A1)*Y
1950 IF X<=H-1E-5 THEN 1870
1960 IF ABS(X-H)<1E-5 THEN 2140
1970 GOSUB 2340
1980 IF X>=Y THEN 2180
1990 B1=180-B
2000 C1=180-A-B1
2010 Z1=X*SIN(C1*P1)/SIN(A1)
2020 PRINT
2030 PRINT
2040 PRINT "THERE ARE TWO TRIANGLES:"
2050 PRINT
2060 PRINT TAB(5);"TRIANGLE 1:";TAB(47);"TRIANGLE 2:"
2070 PRINT "ANGLES","SIDES"," ","ANGLES","SIDES"
2080 PRINT "(DEGREES)","(UNITS)"," ","(DEGREES)","(UNITS)"
2090 PRINT
2100 PRINT "A=",A,"X=",X," ","A=",A,"X=",X
2110 PRINT "B=",B,"Y=",Y," ","B=",B1,"Y=",Y
2120 PRINT "C=",C,"Z=",Z," ","C=",C1,"Z=",Z1
2130 GOTO 2270
2140 Z=Y*COS(A1)
2150 B=90
2160 C=90-A
2170 REM---------PRINT-OUT----------
2180 PRINT
2190 PRINT
2200 PRINT "ANGLES","SIDES"
2210 PRINT "(DEGREES)","(UNITS)"
2220 PRINT
2230 PRINT "A=",A,"X=",X
2240 PRINT "B=",B,"Y=",Y
2250 PRINT "C=",C,"Z=",Z
2260 REM---,------REPEAT?----------
2270 PRINT
2280 PRINT
2290 PRINT "DO YOU WISH TO RUN AGAIN";
2300 INPUT R$
2310 IF R$>="Y" THEN 1190
2320 STOP
2330 REM---------SUBROUTINE FOR SSA----------
2340 B=FNS(Y*SIN(A1)/X)
2350 C=180-A-B
2360 Z=X*SIN(C*P1)/SIN(A1)
2370 RETURN
2380 REM---------ARCSIN AND ARCCOS FUNCTIONS----------
2390 DEF FNS(Y)=ATN(Y/SQR(1-Y↑2))/P1        'ARCSIN
2400 DEF FNC(Y)=90-FNS(Y)                   'ARCCOS
2410 END
```

OUTPUT:

THIS PROGRAM FINDS THE REMAINING PARTS OF ANY TRIANGLE,
GIVEN A SIDE AND ANY TWO OTHER PARTS.

7-2 Program (11): Solutions of Triangles in Trigonometry

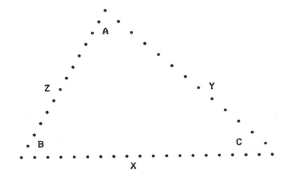

```
ARE YOU GIVEN ASA, SAS, SSS, OR SSA ?ASA

WHAT IS THE MEASURE OF ANGLE A IN DEGREES ?30
WHAT IS THE MEASURE OF ANGLE B IN DEGREES ?19
WHAT IS THE LENGTH OF SIDE Z ?12


ANGLES          SIDES
(DEGREES)       (UNITS)

A= 30           X= 7.95006
B= 19           Y= 5.17657
C= 131          Z= 12


DO YOU WISH TO RUN AGAIN ?YES

ARE YOU GIVEN ASA, SAS, SSS, OR SSA ?SAS

WHAT IS THE LENGTH OF SIDE Y ?12
WHAT IS THE LENGTH OF SIDE Z ?5.2
WHAT IS THE MEASURE OF ANGLE A IN DEGREES ?30


ANGLES          SIDES
(DEGREES)       (UNITS)

A= 30           X= 7.93473
B= 130.872      Y= 12
C= 19.1276      Z= 5.2


DO YOU WISH TO RUN AGAIN ?YES

ARE YOU GIVEN ASA, SAS, SSS, OR SSA ?SSS
```

```
WHAT IS THE LENGTH OF SIDE X ?8
WHAT IS THE LENGTH OF SIDE Y ?12
WHAT IS THE LENGTH OF SIDE Z ?5.2


ANGLES          SIDES
(DEGREES)       (UNITS)

A= 30.9415      X= 8
B= 129.534      Y= 12
C= 19.5243      Z= 5.2


DO YOU WISH TO RUN AGAIN ?YES


ARE YOU GIVEN ASA, SAS, SSS, OR SSA ?SSA


WHAT IS THE LENGTH OF SIDE X ?8
WHAT IS THE LENGTH OF SIDE Y ?12
WHAT IS THE MEASURE OF ANGLE A IN DEGREES ?30


THERE ARE TWO TRIANGLES:
       TRIANGLE 1:                              TRIANGLE 2:
ANGLES          SIDES                    ANGLES          SIDES
(DEGREES)       (UNITS)                  (DEGREES)       (UNITS)

A= 30           X= 8                     A= 30           X= 8
B= 48.5904      Y= 12                    B= 131.41       Y= 12
C= 101.41       Z= 15.6838               C= 18.5904      Z= 5.1008


DO YOU WISH TO RUN AGAIN ?YES


ARE YOU GIVEN ASA, SAS, SSS, OR SSA ?SSA


WHAT IS THE LENGTH OF SIDE X ?12
WHAT IS THE LENGTH OF SIDE Y ?8
WHAT IS THE MEASURE OF ANGLE A IN DEGREES ?30


ANGLES          SIDES
(DEGREES)       (UNITS)

A= 30           X= 12
B= 19.4712      Y= 8
C= 130.529      Z= 18.242


DO YOU WISH TO RUN AGAIN ?YES


ARE YOU GIVEN ASA, SAS, SSS, OR SSA ?SSA


WHAT IS THE LENGTH OF SIDE X ?1
WHAT IS THE LENGTH OF SIDE Y ?2
WHAT IS THE MEASURE OF ANGLE A IN DEGREES ?:
```

7-2 Program (11): Solutions of Triangles in Trigonometry

```
ANGLES           SIDES
(DEGREES)        (UNITS)

A= 30            X= 1
B= 90            Y= 2
C= 60            Z= 1.73205


DO YOU WISH TO RUN AGAIN ?YES


ARE YOU GIVEN ASA, SAS, SSS, OR SSA ?SSS


WHAT IS THE LENGTH OF SIDE X ?5
WHAT IS THE LENGTH OF SIDE Y ?6
WHAT IS THE LENGTH OF SIDE Z ?1


NO TRIANGLE SATISFIES YOUR VALUES OF X, Y, AND Z.


DO YOU WISH TO RUN AGAIN ?YES


ARE YOU GIVEN ASA, SAS, SSS, OR SSA ?SSA


WHAT IS THE LENGTH OF SIDE X ?2
WHAT IS THE LENGTH OF SIDE Y ?5
WHAT IS THE MEASURE OF ANGLE A IN DEGREES ?30


NO TRIANGLE SATISFIES YOUR VALUES OF X, Y, AND A.


DO YOU WISH TO RUN AGAIN ?NO
```

---

7. Comments about Program (11)

a. Note the organization of the program. The REM statements separate the program into various parts: DIAGRAM, CASE, ASA, SAS, SSS, SSA, PRINT-OUT, and REPEAT. There is also a special subroutine for SSA and definitions of the ARCSIN and ARCCOS in terms of the ATN function.

Although this program is somewhat lengthy, the routines in each part are relatively short and almost independent of the other parts; thus the main program consists of a number of shorter programs.

b. Study very carefully the technique of the DIAGRAM routine. Notice the heavy reliance on the TAB function. In Steps 1060 and 1070, if $I = 12$ or $I = 7$, then the computer respectively jumps to either Step 1120 or 1140 to label the diagram.

c. Since angles are entered and typed in degree measure, and the computer uses radian measure, we must convert from degrees to radians and from radians to degrees in the program. In Step 1210 we set P1 = 3.14159/180 to help us use the conversion formula:

$$m^R(\beta) = \frac{\pi}{180} \cdot m^\circ(\beta) \ .$$

d. We have defined the Arcsin and Arccos functions in Steps 2390 and 2400 according to Equation (4) of Paragraph 2 and Equation (8) of Paragraph 3 respectively. Notice in Step 2390 that we divided ATN(Y/SQR(1-Y↑2)) by P1. Dividing by P1 in this step converts the value of FNS to a degree measure. Since the value of FNS is in degree measure, then so is the value of FNC in Step 2400.

Also in Step 1520, in order to set B equal to Arcsin (D), we simply used the statement B = FNS(D), because FNS is the name for Arcsin. Since FNS(D) is measured in degrees, then B is also measured in degrees. Similarly, in Step 1700, in order to set A equal to Arccos (D), we used the statement A = FNC(D), because FNC is the name for Arccos. Note that FNC is also measured in degrees.

e. The ambiguous case SSA given in Steps 1770 through 2160 is a rather lengthy routine which should be carefully studied.

The statement in Step 1960 is quite significant, in that this is the case for which X = H, where H = Y · Sin (A). However, if we simply wrote IF X = H THEN 2140 in Step 1960, then, because of the round-off error of the computer, the possibility exists that the machine would not give us the correct answer.

For example, if X = $\sqrt{3}$, Y = 2 and A = 60°, then we would have one right triangle ABC, in which C = 30°, B = 90° and Z = 1. However, the computer would approximate both $\sqrt{3}$ and 2 · Sin 60°. If the two approximations were not exactly equal, the computer would give us the results for two triangles or no triangle, dependent on whether the round-off value for H is less than or greater than the round-off value for X.

In order to avoid this situation, we allow for a difference of 1E-5 between X and H in Steps 1950 and 1960.

EXERCISES 7-2

1. Write a program which demands two sides and the included angle of any triangle ABC, and then computes and types the area of triangle ABC.

2. Write a program which demands the three sides of any triangle ABC, and then computes and types the area of the triangle ABC.

3. Write a program which demands any three parts of a triangle ABC, one of which must be a side, and then computes and types the area of triangle ABC.

4. Write a program which prints a diagram of a square. Write the program so that the computer labels each side and angle.

5. Write a program which prints a diagram of a diamond-shaped figure. Write the program so that the computer labels each side and angle.

7-3 Program (12): The Graph of Any Function

1. Rectangular Coordinate System

Near the end of your first-year algebra course, you learned that a function f is a set of ordered pairs (x, y) such that for a given value of x from a domain of f there exists one and only one value of y in the range of f. You learned that you could specify a function f by a rule or equation such as $f(x) = ax^2 + bx + c$, $a \neq 0$.

You learned how to graph a function. You know that for a particular value of x you can find the corresponding value of f(x). You can even use the computer to find f(x), as shown in Example (1) of Section 6-4.

In order to construct the graph of a function, you set up a rectangular coordinate system. By convention, the values of x are measured on the horizontal axis, and the values of y are taken on the vertical axis, as shown in Figure (15).

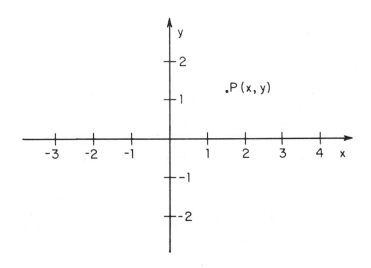

Figure (15)

A convenient unit of measure is used along the x-axis, and the *same* unit of measure is taken along the y-axis. For any ordered pair (x, y) of real numbers which belong to a function f, there exists a point P on the graph of f. Also, for any point P on the graph of a function f, there is one ordered pair (x, y) of real numbers which satisfies the equation of f.

For example, a sketch of the graph of the function $f(x) = x^2 - 2x - 3$ is given in Figure (16).

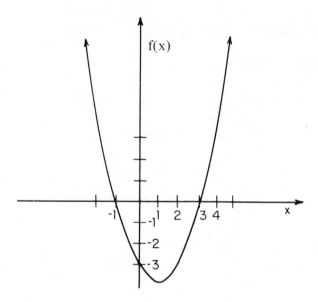

Figure (16)

There are a number of methods you could use to sketch the graph of Figure (16), but one of the simplest (and probably the most tedious) is to find many ordered pairs which satisfy the equation $f(x) = x^2 - 2x - 3$, plot those points on a coordinate system, and then draw a smooth curve through those points. Of course, we are somewhat familiar with the characteristic shape of the graph of an equation of the form $f(x) = ax^2 + bx + c$, where $a \neq 0$. We know that its graph is a parabola. If an equation of a function were somewhat complicated, such as $f(x) = 2^{\sin x} - 1$, or $f(x) = \cos(2x) + \log_e(x) + .5$, then the problem of constructing a rather accurate graph becomes much more difficult.

2. Graph by Computer

The computer can graph the latter two functions of Paragraph 1 above just as easily as the first function! The procedure we shall use on the computer is to calculate the value of F(X) for any given value of X, and to have the computer type a period (.) to indicate a point P as the graph of the ordered pair (X, F(X)).

There is one convention we must change when we have the computer graph a function for us: the X-axis is vertical and the Y-axis is horizontal. This change of coordinates greatly simplifies the problem of plotting a point on the computer.

If we simply rotate the conventional axis 90° in a clockwise direction, we will have the coordinate system used by the computer, as you can see in Figure (17).

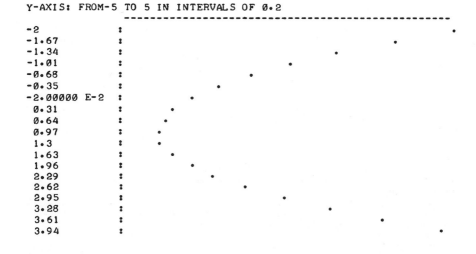

Figure (17)

After the computer completes its sketch, which consists of nothing but dots, you can easily make a smooth curve through the dots, and include a coordinate axes, as shown in Figure (18).

The basic idea of the program is to compute the value of F(X) for a given value of X, and to direct the computer to space across the page to the proper position and type a period. The proper position is determined by the value of F(X) within the range of the graph.

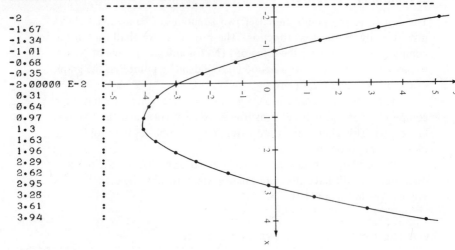

Figure (18)

3. Program (12)

The program is as follows.

PROGRAM (12):

```
10 DEF FNA(X)=X
20 PRINT "THIS PROGRAM PLOTS ANY FUNCTION."
30 PRINT "HAVE YOU DEFINED YOUR FUNCTION";
40 INPUT Q$
50 IF Q$>="Y" THEN 90
60 PRINT "TYPE '10 DEF FNA(X)=(YOUR FUNCTION)', HIT <RETURN> KEY,"
70 PRINT "THEN TYPE 'RUN', <RETURN>."
80 STOP
90 PRINT
100 PRINT
110 PRINT "WHAT IS YOUR MINIMUM VALUE OF X";
120 INPUT A
130 PRINT "WHAT IS YOUR MAXIMUM VALUE OF X";
140 INPUT B
150 PRINT "HOW LONG ARE YOUR SUBINTERVALS ON THE X-AXIS";
160 INPUT R
170 PRINT "WHAT IS YOUR MINIMUM VALUE OF Y";
180 INPUT C
190 PRINT "WHAT IS YOUR MAXIMUM VALUE OF Y";
200 INPUT D
210 L=(D-C)/50
220 C1=C-L/2
230 D1=D+L/2
240 PRINT
250 PRINT
260 PRINT
270 PRINT "Y-AXIS: FROM"C;"TO"D;"IN INTERVALS OF"L
280 PRINT TAB(15);"-----------------------------------------------------"
```

```
290 FOR X=A TO B STEP R
300    Y=FNA(X)
310    IF Y<C1 THEN 380
320    IF Y>=D1 THEN 380
330    PRINT X;TAB(14);":";TAB((Y-C)/L+15.5);"."
340 NEXT X
350 PRINT
360 PRINT
370 STOP
380 PRINT X;TAB(14);":"
390 GOTO 340
400 END
```

OUTPUT (1):

```
THIS PROGRAM PLOTS ANY FUNCTION.
HAVE YOU DEFINED YOUR FUNCTION ?NO
TYPE '10 DEF FNA(X)=(YOUR FUNCTION)', HIT <RETURN> KEY,
THEN TYPE 'RUN', <RETURN>.



 10 DEF FNA(X)=X↑2-2*X-3
 RUN



THIS PROGRAM PLOTS ANY FUNCTION.
HAVE YOU DEFINED YOUR FUNCTION ?YES


WHAT IS YOUR MINIMUM VALUE OF X ?-2
WHAT IS YOUR MAXIMUM VALUE OF X ?4
HOW LONG ARE YOUR SUBINTERVALS ON THE X-AXIS ?.33
WHAT IS YOUR MINIMUM VALUE OF Y ?-5
WHAT IS YOUR MAXIMUM VALUE OF Y ?5


Y-AXIS: FROM-5 TO 5 IN INTERVALS OF 0.2
               -------------------------------------------------
 -2            :                                                .
 -1.67         :                                         .
 -1.34         :                                    .
 -1.01         :                              .
 -0.68         :                         .
 -0.35         :                    .
 -2.00000 E-2  :               .
  0.31         :           .
  0.64         :          .
  0.97         :         .
  1.3          :         .
  1.63         :          .
  1.96         :             .
  2.29         :                .
  2.62         :                    .
  2.95         :                        .
  3.28         :                             .
  3.61         :                                   .
  3.94         :                                          .
```

176   Computer Programming in BASIC

4. Comments About Program (12)

a. Notice the organization of the program. Steps 10 through 70 are concerned with defining your function; Steps 110 through 200 ask for the limiting values of X and Y and the increment value along the X-axis; Step 210 determines the length of each space along the Y-axis; Steps 270 and 280 print and label the Y-axis; and Steps 290 through 390 plot the function.

b. The INPUT statements in Steps 110 through 200 ask for the lower and upper limits of the X-value, the increment value of X, and the lower and upper limits of the Y-value.

You can easily arrange to have the unit of measure along the X and Y axes to be almost the same by a little arithmetic before you start to execute the program. For example, for the equation $F(X) = X^2 - 2X - 3$, if $X = 0$, then $F(X) = -3$. The graph will probably have a value less than -3, since F is not symmetric with respect to the Y-axis. Let us then let the values of Y range between -5 and 5. We have designed the program so that there are 50 carriage spaces available for the value of $F(X)$. The unit of measure along the Y-axis is then $(5-(-5))/50 = 0.2$.

Since there are 6 lines per inch and 10 spaces per inch on the Teletype, let us have the increment value of X be such that the ratio

$$\frac{Y\text{- increment}}{X\text{- increment}} = \frac{6}{10}.$$

Thus

$$\frac{.2}{X\text{- increment}} = \frac{3}{5}$$

and

$$X\text{- increment} = 0.33 .$$

Let us also agree to let the lower limit of X to be -2 and the upper limit 4. The resulting graph is shown in the OUTPUT(1) of Program (12) above.

c. In Step 220, C1 is assigned the value $C-L/2$. Step 310 in the loop of Steps 290 through 340 determines whether $F(X)$ is greater than the minimum value C assigned to the Y-axis. Any value of $F(X)$ which differs from C by one-half the length of a space will be printed in the first space of the graph, including values of $F(X)$ slightly less than C.

Similarly, Step 230 sets D1 equal to $D+L/2$. Step 320 then determines whether or not $F(X)$ is less than the maximum value D assigned to the Y-axis. The relation $>=$ was used in Step 320 rather than simply $>$ so that the computer would not print a period in the first space to the right of the value of D, since that space is outside the range of F(X). Steps 310 and 320 thus ensure that the value of F(X) lies within the prescribed lower and upper limits on the Y-axis.

d. If a given value of F(X) lies within the prescribed limits on the Y-axis, then the computer continues to Step 330. The first part of the statement in Step 330 prints the value of X, and then types a colon in the 15th space. The statement TAB ((Y−C)/L+15.5) : (1) determines the difference between the minimum value C of the Y-axis and the value of Y; (2) divides this number by L in order to obtain the proper number of spaces between the first space of the graph and the space in which the value of F(X) should be plotted; (3) adds 15.5 to this value (the number 15 is used here because the value of X and the colon require 15 spaces, and the number 0.5 is used in order to round off properly the argument of the TAB function); (4) takes the INT of this number (recall that the TAB function has a built-in INT function); (5) prints the TAB of this value by moving the carriage across the page to the proper space; (6) and then prints a period as the point on the graph of F.

e. It is easy enough for us to make a rather accurate graph of some of the more common functions with equations such as $F(X) = X^2 - 2X - 3$ without the computer.

Now consider a function whose equation is $F(X) = COS(2X) + LOG(X) + 0.5$. We would probably have some difficulty in graphing this function "by hand." However, the computer can make the graph of this function just as easily as the graph of $F(X) = X^2 - 2X - 3$, as follows.

```
OUTPUT (2):

THIS PROGRAM PLOTS ANY FUNCTION.
HAVE YOU DEFINED YOUR FUNCTION ?NO
TYPE '10 DEF FNA(X)=(YOUR FUNCTION)', HIT <RETURN> KEY,
THEN TYPE 'RUN', <RETURN>.


10 DEF FNA(X)=COS(2*X)+LOG(X)+0.5
RUN


THIS PROGRAM PLOTS ANY FUNCTION.
HAVE YOU DEFINED YOUR FUNCTION ?YES

WHAT IS YOUR MINIMUM VALUE OF X ?.1
WHAT IS YOUR MAXIMUM VALUE OF X ?5.5
HOW LONG ARE YOUR SUBINTERVALS ON THE X-AXIS ?.18
WHAT IS YOUR MINIMUM VALUE OF Y ?-1
WHAT IS YOUR MAXIMUM VALUE OF Y ?4
```

Y-AXIS: FROM-1 TO 4 IN INTERVALS OF 0.1

 0.1   :   •
 0.28   :   •
 0.46   :   •
 0.64   :   •
 0.82   :   •
 1   :   •
 1.18   :   •
 1.36   :   •
 1.54   :   •
 1.72   :   •
 1.9   :   •
 2.08   :   •
 2.26   :   •
 2.44   :   •
 2.62   :   •
 2.8   :   •
 2.98   :   •
 3.16   :   •
 3.34   :   •
 3.52   :   •
 3.7   :   •
 3.88   :   •
 4.06   :   •
 4.24   :   •
 4.42   :   •
 4.6   :   •
 4.78   :   •
 4.96   :   •
 5.14   :   •
 5.32   :   •

f. An interesting and extremely useful function in statistics is the normal distribution function, whose equation is

$$F(X) = \frac{1}{\sqrt{2\pi}} \cdot e^{-X^2/2}.$$

The INPUT data and resulting graph of the normal distribution function is as follows.

OUTPUT (3):

```
THIS PROGRAM PLOTS ANY FUNCTION.
HAVE YOU DEFINED YOUR FUNCTION ?NO
TYPE '10 DEF FNA(X)=(YOUR FUNCTION)', HIT <RETURN> KEY,
THEN TYPE 'RUN', <RETURN>.


10 DEF FNA(X)=EXP(-X↑2/2)/SQR(2*3.14159)
RUN



THIS PROGRAM PLOTS ANY FUNCTION.
HAVE YOU DEFINED YOUR FUNCTION ?YES
```

7-3 Program (12): The Graph of Any Function 179

```
WHAT IS YOUR MINIMUM VALUE OF X ?-3
WHAT IS YOUR MAXIMUM VALUE OF X ?3
HOW LONG ARE YOUR SUBINTERVALS ON THE X-AXIS ?.111111
WHAT IS YOUR MINIMUM VALUE OF Y ?0
WHAT IS YOUR MAXIMUM VALUE OF Y ?.45


Y-AXIS: FROM 0 TO 0.45 IN INTERVALS OF 9.00000 E-3
       ---------------------------------------------------
-3              :.
-2.88889        : .
-2.77778        : .
-2.66667        : .
-2.55556        :  .
-2.44445        :  .
-2.33333        :   .
-2.22222        :    .
-2.11111        :     .
-2.             :      .
-1.88889        :       .
-1.77778        :        .
-1.66667        :          .
-1.55556        :            .
-1.44445        :              .
-1.33334        :                 .
-1.22222        :                     .
-1.11111        :                         .
-1.             :                             .
-0.888891       :                                  .
-0.77778        :                                       .
-0.666669       :                                            .
-0.555558       :                                                .
-0.444447       :                                                    .
-0.333336       :                                                       .
-0.222225       :                                                         .
-0.111114       :                                                          .
-3.12924 E-6    :                                                           .
 0.111108       :                                                          .
 0.222219       :                                                         .
 0.33333        :                                                       .
 0.444441       :                                                    .
 0.555552       :                                                .
 0.666663       :                                            .
 0.777774       :                                       .
 0.888885       :                                  .
 0.999996       :                             .
 1.11111        :                         .
 1.22222        :                     .
 1.33333        :                 .
 1.44444        :              .
 1.55555        :            .
 1.66666        :          .
 1.77777        :        .
 1.88888        :       .
 1.99999        :      .
 2.11111        :     .
 2.22222        :    .
 2.33333        :   .
 2.44444        :  .
 2.55555        :  .
 2.66666        : .
 2.77777        : .
 2.88888        : .
 2.99999        :.
```

EXERCISES 7-3

1. Write a program which plots any function. The program should ask for a domain and suggest a range and increment value along the X-axis.
2. Write a program which plots two functions at the same time.
3. Write a program which plots any function, but use the CHANGE command. (Hint: Set variables P(1) to P(55) equal to 32, the ASCII code number for a space. Then set the appropriate variable P(X) equal to 46, the code number for a period.)
4. Write a program which plots up to ten functions at the same time.

7-4 Program (13): Real Zeros of Any Function

1. Zeros of a Function

You will recall from your work in mathematics that the zeros of a function F are the roots of the corresponding equation $F(X) = 0$. For example, the zeros of $F(X) = X^2 - 2X - 3$ are the roots of the equation $X^2 - 2X - 3 = 0$. We have:

$$X^2 - 2X - 3 = 0$$

$$(X-3)(X+1) = 0$$

$$X = 3 \text{ or } X = -1.$$

Since $X = 3$ and $X = -1$ will both satisfy the equation $X^2 - 2X - 3 = 0$, then the zeros of $F(X) = X^2 - 2X - 3$ are 3 and -1.

You have already seen the graph of $F(X) = X^2 - 2X - 3$ in Figure (16) above. Notice on the graph in Figure (16) that the function F intersects the X-axis at $X = 3$ and $X = -1$.

2. Continuous Function

We say that the function $F(X) = X^2 - 2X - 3$ is *continuous* if the graph of F consists of an unbroken curve, as in Figure (16). (The formal definition of a continuous function is beyond the scope of this book.)

If a function is continuous, then we shall assume that if $F(A)$ and $F(B)$ are of opposite sign, then there is at least one real number $X = C$ where $A < C < B$, such that $F(C) = 0$. For example, if $F(X) = X^2 - 2X - 3$ and if $A = -2$ and $B = 0$, then $F(-2) = 5$ and $F(0) = -3$. Hence there is at least one real number $X = C$, where $-2 < C < 0$, such that $F(C) = 0$.

From Figure (16), you can see that C = -1. Similarly, F(2) = -3 and F(4) = 5. Therefore, there exists at least one real number C, where 2 < C < 4, such that F(C) = 0. Again from Figure (16), you see that C = 3.

3. Finding Zeros by the Computer.

Since the computer can very easily calculate values of F(X) for any function F and for any real number X, then we can write a program which divides an interval from A to B into N subintervals as follows: A = X(0), X(1), X(2), ..., X(I), ..., X(N) = B. We can have the computer compare the signs of F(A) and F(X(I)), as X(I) goes from A to B, until it finds a change of sign. If there is a change of sign, the computer can then "search" that interval from X(I-1) to X(I) for some number C such that F(C) = 0. The computer will thus have found one zero C of the function F.

After the computer has found a zero C, it can then search the interval from C to B for another change of sign, etc. In this way, the computer can find *all* the real zeros of a function F from A to B.

4. Program (13)

The program for finding all the real zeros of any function is as follows.

PROGRAM (13):

```
100 DEF FNC(X)=X
110 PRINT "THIS PROGRAM FINDS THE REAL ZEROS OF A FUNCTION."
120 PRINT
130 REM---------DEFINING FUNCTION----------
140 PRINT "HAVE YOU DEFINED YOUR FUNCTION ";
150 INPUT R$
160 IF R$>="Y" THEN 190
170 PRINT "TYPE '100 DEF FNC(X)=(YOUR FUNCTION)',<RETURN>,'RUN'."
180 STOP
190 PRINT
200 PRINT
210 PRINT "WHAT IS THE LOWER BOUND OF YOUR INTERVAL ";
220 INPUT A1
230 PRINT "WHAT IS THE UPPER BOUND OF YOUR INTERVAL ";
240 INPUT B
250 PRINT
260 PRINT
270 PRINT
280 PRINT
290 LET A=A1
300 LET N=0
310 REM---------CHANGE OF SIGN----------
320 LET Q=SGN(FNC(A))
330 LET H=(B-A)/100
```

```
340 FOR X=A TO B STEP H
350     LET S=FNC(X)
360     IF S=0 THEN 510
370     IF C*S<0 THEN 430
380 NEXT X
390 IF N=1 THEN 580
400 PRINT "THERE ARE NO REAL ROOTS IN YOUR INTERVAL."
410 GOTO 590
420 REM---------SEARCH FOR ZERO----------
430 LET X(1+0)=X-H
440 LET X(1-0)=X
450 LET X=(X(0)+X(2))/2
460 LET G=SGN(FNC(X))
470 IF G=0 THEN 510
480 LET X(1+G)=X
490 IF ABS(X(0)-X(2))/(ABS(X(0))+ABS(X(2)))>=1E-6 THEN 450
500 REM---------PRINT ZERO----------
510 IF N=1 THEN 540
520 PRINT "THE REAL ZEROS FROM"A;"TO"B;"OF YOUR FUNCTION ARE:"
530 LET N=1
540 PRINT X;
550 LET A=X+1E-5
560 GOTO 320
570 REM---------REPEAT?----------
580 PRINT
590 PRINT
600 PRINT
610 PRINT
620 PRINT "DO YOU WISH TO RUN AGAIN ";
630 INPUT R$
640 IF R$>="Y" THEN 120
650 END
```

OUTPUT (1):

THIS PROGRAM FINDS THE REAL ZEROS OF A FUNCTION.

HAVE YOU DEFINED YOUR FUNCTION ?NO
TYPE '100 DEF FNC(X)=(YOUR FUNCTION)',<RETURN>,'RUN'.

100 DEF FNC(X)=X↑2-2*X-3
RUN

THIS PROGRAM FINDS THE REAL ZEROS OF A FUNCTION.

HAVE YOU DEFINED YOUR FUNCTION ?YES

WHAT IS THE LOWER BOUND OF YOUR INTERVAL ?-10
WHAT IS THE UPPER BOUND OF YOUR INTERVAL ?10

THE REAL ZEROS FROM-10 TO 10 OF YOUR FUNCTION ARE:
-1.   3.

7-4 Program (13): Real Zeros of Any Function   183

```
DO YOU WISH TO RUN AGAIN ?YES

HAVE YOU DEFINED YOUR FUNCTION ?NO
TYPE '100 DEF FNC(X)=(YOUR FUNCTION)',<RETURN>,'RUN'.



100 DEF FNC(X)=X↑2+1
RUN



THIS PROGRAM FINDS THE REAL ZEROS OF A FUNCTION.

HAVE YOU DEFINED YOUR FUNCTION ?YES

WHAT IS THE LOWER BOUND OF YOUR INTERVAL ?-25
WHAT IS THE UPPER BOUND OF YOUR INTERVAL ?25


THERE ARE NO REAL ROOTS IN YOUR INTERVAL.


DO YOU WISH TO RUN AGAIN ?NO
```

---

5. Comments about Program (13)

a. The REM statements again separate the program into parts. Steps 100 to 280 describe the program, explain how to define your function, and ask for the lower and upper bounds of the interval to be searched. Steps 320 to 380 search the interval for a change of sign in F. Steps 430 to 490 find the zero in that interval in which the change of sign was found, and Steps 510 to 540 print the zeros. Steps 580 to 640 ask if you wish to repeat the program.

b. The loop in Steps 340 to 380 searches for a change in sign. For the first search, $A = A1$ (Step 290). In Step 320, Q is set equal to SGN(FNC(A)). The interval from A to B is divided into 100 subintervals, and each value of $S = FNC(X)$ is computed.

If $S = 0$, then the computer has found a zero and jumps to Step 510, since the condition of Step 360 is satisfied. The zero is printed (Step 540, A is redefined (Step 550), and the computer jumps back to Step 320. The computer then searches this new interval for another change of sign.

If $S <> 0$ in Step 360, then the computer continues to the next step and compares the signs of S and Q. If they are of the same sign, then Q∗S > 0 and the computer continues to Step 380. If Q∗S < 0, then there is a change of sign in that interval. The computer "knows" that this change of sign occurs between X and X-H, where H = (B-A)/100, since FNC(X-H) has the same sign as Q, and FNC(X) has the opposite sign of Q. Thus the computer jumps to Step 430 and searches for a zero between X-H and X.

c. Step 430 assigns the value X-H to X(1+Q), where Q is either -1 or 1. Thus, X(1+Q) = X(0) or X(2). Similarly, X(1-Q) is assigned the value X. Now, X(1-Q) equals either X(0) or X(2), whichever one X(1+Q) does not equal; i.e., if X(0) = X-H then X(2) = X, and if X(0) = X, then X(2) = X-H.

In Step 450, X is assigned the average value of X-H and X. Now, if the sign G of FNC(X) is 0, then the computer has found a zero and goes to Step 510. If G <> 0, then G = 1 or G = -1. Thus, in Step 480, X(1+G) = X(0) = X or X(1+G) = X(2) = X, where X is the average value determined in Step 450.

In either case, the computer has divided the interval containing a change of sign in half, and called the midpoint value X.

Let us call this midpoint value a new variable D in the discussion that follows.

Either D = X(1+G) = X(2) or D = X(1+G) = X(0). If G has the same sign as FNC(X(0)), then D = X(0). If G has the same sign as FNC(X(2)), then D = X(2).

Thus, when the computer jumps back to Step 450 from Step 490, it computes a new average value of X (which we shall call E), using either

$$E = \frac{(X-H) + D}{2} \quad \text{or} \quad E = \frac{D + X}{2},$$

dependent on whether FNC(D) is of the same sign as the original FNC(X) (in that case, $E = \frac{(X-H) + D}{2}$) or FNC(D) is of the same sign as FNC(X-H) (then $E = \frac{D + X}{2}$). The figure on the next page illustrates this latter situation.

The computer continues through the loop of Steps 450 through 490 until G = 0 in Step 470 or until the relative error between X(0) and X(2) is less than 1E-6, as given in Step 490. As soon as one of these two conditions is satisfied, the computer prints X, redefines the beginning of the interval A (Step 550), and returns to Step 320 to search the next interval for a change in sign.

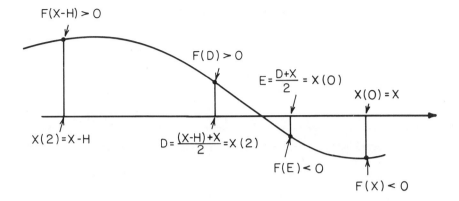

Figure (19)

d. In Step 490 the expression

$$\frac{|A-B|}{|A| + |B|}$$

determines the difference between A and B relative to the size of A and B. Thus, we are assured of having accuracy to six significant digits. The reason why a statement such as IF ABS (A-B) >= 1E-6 THEN 450 is unsatisfactory is that, for a small zero such as 0.0001, we would have only three significant digits. Also, a large zero such as 1000000 would require 13 significant digits, a number beyond the capability of the computer.

e. It is virtually impossible to write a program for a zero-search of a function which will handle all situations. The possibility exists, particularly for functions with zeros which are very close together, that the technique used in this program will not find them. Also, this program will not find the zeros of a function whose graph just touches the X-axis, such as $F(X) = X^2$, unless the zero happened by chance to be one of the 100 subdivisions.

If 100 subdivisions are insufficient to find two closely packed zeros, then one could use a larger number. However, bear in mind that the Central Processing Time increases as the number of subintervals increases.

f. To find the zeros of the function $F(X) = X^2 - 2X - 3$ using the computer is probably a waste of computer time, since we can find the zeros "by hand." However, suppose we were asked to find the zeros of the function whose graph is given in OUTPUT (2) of Program (12) above: $F(X) = COS(2X) + LOG(X) + 0.5$.

The zeros of this function are quite difficult to find by hand. Using Program (13), the problem becomes quite simple, as you can see in the following OUTPUT (2).

OUTPUT (2):

```
THIS PROGRAM FINDS THE REAL ZEROS OF A FUNCTION.

HAVE YOU DEFINED YOUR FUNCTION ?NO
TYPE '100 DEF FNC(X)=(YOUR FUNCTION)',<RETURN>,'RUN'.


100 DEF FNC(X)=COS(2*X)+LOG(X)+0.5
RUN


THIS PROGRAM FINDS THE REAL ZEROS OF A FUNCTION.

HAVE YOU DEFINED YOUR FUNCTION ?YES


WHAT IS THE LOWER BOUND OF YOUR INTERVAL ?0.001
WHAT IS THE UPPER BOUND OF YOUR INTERVAL ?20



THE REAL ZEROS FROM 0.001 TO 20 OF YOUR FUNCTION ARE:
 0.252891   1.10972   1.63514



DO YOU WISH TO RUN AGAIN ?NO
```

EXERCISES 7-4

1. Write a program which finds one root in an interval for an equation $F(X) = 0$. Use a random variable method to search for a change in sign.

2. Write a program which uses Newton's method to approximate the zeros of a function.

3. Write a program which systematically searches for an interval in which a change of sign of the function occurs, then searches this interval for a smaller interval in which a change of sign occurs, etc.,

until an interval is found whose endpoints differ by a relative error less than 1E-6.

4. Write a program which asks for an unspecified number of words (use a MAT INPUT statement with a string variable) and then types the words in alphabetical order.

7-5 Summary

You saw the real power of the computer used in this chapter. After some discussion about the techniques of debugging and a general approach to the writing of programs, you then learned how to write three rather significant programs for problems with which you deal for the first time on the secondary school level: (1) to find the solution of triangles in trigonometry; (2) to graph any well-defined function; and (3) to find the real zeros of any continuous function. These latter two programs are useful not only to the secondary-school student of mathematics but indeed to anyone who works with functions: the college and graduate-school student, the engineer, and the applied and theoretical mathematician.

CHAPTER 8

In this chapter you will first review some of the fundamental concepts of matrices. You will also learn how to use the MATrix statements of the BASIC language. You will then solve a system of n linear equations each with n variables by means of matrices and by using the MAT statements on the computer.

8-1 Matrices

1. What Is a Matrix?

A rectangular array of numbers such as

$$E = \begin{pmatrix} 3 & 2 & 4 \\ 5 & 1 & 2 \end{pmatrix}$$

is called a matrix. Each number in the array is called an *entry* of the matrix. We call E a 2 x 3 matrix because it contains 2 rows and 3 columns. The entries in the first row are 3, 2, and 4. Those in the second row are 5, 1, and 2. The entries in the first column are 3 and 5; column two contains the numbers 2 and 1, whereas 4 and 2 are the entries in column three. In general, we call a matrix with r rows and c columns an r x c matrix.

The entry in the first row and third column may be written e_{13}. In matrix E above, $e_{13} = 4$. Notice that we write the number of the row first, and the number of the column second.

In general, a 2 x 3 matrix may be written

$$F = \begin{pmatrix} f_{11} & f_{12} & f_{13} \\ f_{21} & f_{22} & f_{23} \end{pmatrix},$$

in which the subscripts of each entry indicate the position of that number in the matrix.

Two matrices A and B are equal if and only if the corresponding entries in the two matrices are equal. That is,

$$\begin{pmatrix} a_{11} & a_{12} & \cdots & a_{1c} \\ a_{21} & a_{22} & \cdots & a_{2c} \\ \vdots \\ a_{r1} & a_{r2} & \cdots & a_{rc} \end{pmatrix} = \begin{pmatrix} b_{11} & b_{12} & \cdots & b_{1c} \\ b_{21} & b_{22} & \cdots & b_{2c} \\ \vdots \\ b_{r1} & b_{r2} & \cdots & b_{rc} \end{pmatrix}$$

if and only if $a_{ij} = b_{ij}$, where $i = 1, 2, 3, \ldots, r$ and $j = 1, 2, 3, \ldots, c$. We say that the matrices A and B given above are of *dimensions* r x c.

2. The Sum and Scalar Product of Matrices

We can add two matrices A and B by adding all of the corresponding entries in A and B. That is,

$$\begin{pmatrix} a_{11} & a_{12} & \cdots & a_{1c} \\ a_{21} & a_{22} & \cdots & a_{2c} \\ \vdots \\ a_{r1} & a_{r2} & \cdots & a_{rc} \end{pmatrix} + \begin{pmatrix} b_{11} & b_{12} & \cdots & b_{1c} \\ b_{21} & b_{22} & \cdots & b_{2c} \\ \vdots \\ b_{r1} & b_{r2} & \cdots & b_{rc} \end{pmatrix} =$$

$$\begin{pmatrix} a_{11}+b_{11} & a_{12}+b_{12} & \cdots & a_{1c}+b_{1c} \\ a_{21}+b_{21} & a_{22}+b_{22} & \cdots & a_{2c}+b_{2c} \\ \vdots \\ a_{r1}+b_{r1} & a_{r2}+b_{r2} & \cdots & a_{rc}+b_{rc} \end{pmatrix}.$$

For example,

$$\begin{pmatrix} 2 & 3 & 4 \\ 5 & 6 & 7 \end{pmatrix} + \begin{pmatrix} -1 & 2 & 5 \\ 0 & -2 & 4 \end{pmatrix} = \begin{pmatrix} 1 & 5 & 9 \\ 5 & 4 & 11 \end{pmatrix}.$$

You realize that we can add only matrices of the same dimensions of both r and c.

We can multiply a given matrix A by a real number D, called a *scalar,* by multiplying each entry of A by D. That is,

$$D \cdot \begin{pmatrix} a_{11} & a_{12} & \cdots & a_{1c} \\ a_{21} & a_{22} & \cdots & a_{2c} \\ \vdots & & & \\ a_{r1} & a_{r2} & \cdots & a_{rc} \end{pmatrix} = \begin{pmatrix} Da_{11} & Da_{12} & \cdots & Da_{1c} \\ Da_{21} & Da_{22} & \cdots & Da_{2c} \\ \vdots & & & \\ Da_{r1} & Da_{r2} & \cdots & Da_{rc} \end{pmatrix}.$$

3. The Zero Matrix, Additive Inverse, and Difference of Matrices

If all the entries of a matrix are the number zero, then we call that matrix the *zero matrix* Z. Thus,

$$Z = \begin{pmatrix} 0 & 0 & \cdots & 0 \\ 0 & 0 & \cdots & 0 \\ \vdots & & & \\ 0 & 0 & \cdots & 0 \end{pmatrix}.$$

We define the additive inverse of a matrix A as that matrix $-A$ such that $A + (-A) = Z$, where Z is the zero matrix. Each entry of $-A$ is the negative of the corresponding entry in A. That is, if

$$A = \begin{pmatrix} a_{11} & a_{12} & \cdots & a_{1c} \\ a_{21} & a_{22} & \cdots & a_{2c} \\ \vdots & & & \\ a_{r1} & a_{r2} & \cdots & a_{rc} \end{pmatrix}, \text{ then } -A = \begin{pmatrix} -a_{11} & -a_{12} & \cdots & -a_{1c} \\ -a_{21} & -a_{22} & \cdots & -a_{2c} \\ \vdots & & & \\ -a_{r1} & -a_{r2} & \cdots & -a_{rc} \end{pmatrix}.$$

We can therefore define the *difference* A−B of two matrices A and B by

$$A-B = A + (-B).$$

For example, if

$$A = \begin{pmatrix} 1 & 2 & 3 \\ 4 & 5 & 6 \end{pmatrix} \text{ and } B = \begin{pmatrix} 0 & -1 & 2 \\ -3 & 1 & 4 \end{pmatrix}, \text{ then}$$

$$A - B = \begin{pmatrix} 1 & 3 & 1 \\ 7 & 4 & 2 \end{pmatrix}.$$

4. The Transpose of a Matrix

If the rows of a matrix B are the same as the columns of matrix A, then each is called the *transpose* of the other. We write:

$$A = B^T \text{ and } B = A^T.$$

Thus, if

$$A = \begin{pmatrix} a_{11} & a_{12} & \cdots & a_{1c} \\ a_{21} & a_{22} & \cdots & a_{2c} \\ \vdots & & & \\ a_{r1} & a_{r2} & \cdots & a_{rc} \end{pmatrix}, \text{ then } A^T = \begin{pmatrix} a_{11} & a_{21} & \cdots & a_{r1} \\ a_{12} & a_{22} & \cdots & a_{r2} \\ \vdots & & & \\ a_{1c} & a_{2c} & \cdots & a_{rc} \end{pmatrix}.$$

For example, if

$$A = \begin{pmatrix} 2 & 7 & 3 \\ 1 & 6 & 8 \end{pmatrix}, \text{ then } A^T = \begin{pmatrix} 2 & 1 \\ 7 & 6 \\ 3 & 8 \end{pmatrix}.$$

Thus, if A has dimensions r x c, then, A^T has dimensions c x r.

5. The Product of Two Matrices

Consider the following scheme indicating the *product* of two matrices, in which one entry at a time is displayed in each case.

$$\begin{pmatrix} \boxed{1} & \boxed{3} \\ -2 & 6 \end{pmatrix} \begin{pmatrix} \boxed{5} & 1 \\ \boxed{0} & -2 \end{pmatrix} = \begin{pmatrix} 1\times5+3\times0 & \\ & \end{pmatrix} = \begin{pmatrix} 5+0 & \\ & \end{pmatrix} = \begin{pmatrix} 5 & \\ & \end{pmatrix}$$

$$\begin{pmatrix} \boxed{1} & \boxed{3} \\ -2 & 6 \end{pmatrix} \begin{pmatrix} 5 & \boxed{1} \\ 0 & \boxed{-2} \end{pmatrix} = \begin{pmatrix} 5 & 1\times1+3(-2) \\ & \end{pmatrix} = \begin{pmatrix} 5 & 1-6 \\ & \end{pmatrix} = \begin{pmatrix} 5 & -5 \\ & \end{pmatrix}$$

$$\begin{pmatrix} 1 & 3 \\ \boxed{-2} & \boxed{6} \end{pmatrix} \begin{pmatrix} \boxed{5} & 1 \\ \boxed{0} & -2 \end{pmatrix} = \begin{pmatrix} 5 & -5 \\ (-2)5+6\times0 & \end{pmatrix} = \begin{pmatrix} 5 & -5 \\ -10+0 & \end{pmatrix} = \begin{pmatrix} 5 & -5 \\ -10 & \end{pmatrix}$$

$$\begin{pmatrix} 1 & 3 \\ \boxed{-2} & \boxed{6} \end{pmatrix} \begin{pmatrix} 5 & \boxed{1} \\ 0 & \boxed{-2} \end{pmatrix} = \begin{pmatrix} 5 & -5 \\ -10 & (-2)1+6(-2) \end{pmatrix} = \begin{pmatrix} 5 & -5 \\ -10 & -2-12 \end{pmatrix} = \begin{pmatrix} 5 & -5 \\ -10 & -14 \end{pmatrix}$$

The product of two 3 x 3 matrices A and B is defined as follows:

$$\begin{pmatrix} a_{11} & a_{12} & a_{13} \\ a_{21} & a_{22} & a_{23} \\ a_{31} & a_{32} & a_{33} \end{pmatrix} \cdot \begin{pmatrix} b_{11} & b_{12} & b_{13} \\ b_{21} & b_{22} & b_{23} \\ b_{31} & b_{32} & b_{33} \end{pmatrix} =$$

$$\begin{pmatrix} a_{11}b_{11}+a_{12}b_{21}+a_{13}b_{31} & a_{11}b_{12}+a_{12}b_{22}+a_{13}b_{32} & a_{11}b_{13}+a_{12}b_{23}+a_{13}b_{33} \\ a_{21}b_{11}+a_{22}b_{21}+a_{23}b_{31} & a_{21}b_{12}+a_{22}b_{22}+a_{23}b_{32} & a_{21}b_{13}+a_{22}b_{23}+a_{23}b_{33} \\ a_{31}b_{11}+a_{32}b_{21}+a_{33}b_{31} & a_{31}b_{12}+a_{32}b_{22}+a_{33}b_{32} & a_{31}b_{13}+a_{32}b_{23}+a_{33}b_{33} \end{pmatrix} .$$

We can easily generalize this definition of the product of two rxc matrices A and B as that matrix C such that each entry c_{ij} of C is the sum of the products of the corresponding entries in the i*th* row of A and the j*th* column of B.

We can multiply two matrices A and B whose dimensions are not the same. However, the matrix A must have the same number of columns as matrix B has rows. The resulting product matrix C then has the number of rows in A and the number of columns in B. That is,

$$A_{rxc} \cdot B_{cxp} = C_{rxp}.$$

6. The Identity Matrix, and the Inverse of a Matrix

The identity matrix I is defined as a square n x n matrix whose entries along the main diagonal from upper left to lower right are the number 1 and whose other entries are 0. Thus,

$$I = \begin{pmatrix} 1 & 0 & 0 & \cdots & 0 \\ 0 & 1 & 0 & \cdots & 0 \\ 0 & 0 & 1 & \cdots & 0 \\ \vdots & & & & \\ 0 & 0 & 0 & \cdots & 1 \end{pmatrix}$$

is the identity matrix of *order n*.

Any n x n matrix A multiplied by the identity matrix I of order n is equal to A. That is,

$$A \cdot I = I \cdot A = A.$$

For a given n x n matrix A, if there is a n x n matrix B such that $A \cdot B = B \cdot A = I$, then B is called the *inverse* matrix of A, and we write:

$$B = A^{-1}. \quad \text{Also, } A = B^{-1}.$$

We shall not discuss the method by which one can find the inverse of a n x n matrix, but we shall develop the general method for a 2 x 2 matrix. Let

$$A = \begin{pmatrix} a_{11} & a_{12} \\ a_{21} & a_{22} \end{pmatrix} \quad \text{and} \quad A^{-1} = \begin{pmatrix} x_{11} & x_{12} \\ x_{21} & x_{22} \end{pmatrix} \quad \text{such that}$$

$A \cdot A^{-1} = A^{-1} \cdot A = I$. That is, such that

$$\begin{pmatrix} a_{11} & a_{12} \\ a_{21} & a_{22} \end{pmatrix} \cdot \begin{pmatrix} x_{11} & x_{12} \\ x_{21} & x_{22} \end{pmatrix} = \begin{pmatrix} 1 & 0 \\ 0 & 1 \end{pmatrix}.$$

If we can find the values of x_{11}, x_{12}, x_{21}, and x_{22}, then we have found the entries of the inverse matrix A^{-1} of A.

Using the definition of the product of two matrices, we have:

$$\left. \begin{aligned} a_{11}x_{11} + a_{12}x_{21} &= 1 \\ a_{21}x_{11} + a_{22}x_{21} &= 0 \end{aligned} \right\} \quad (1) \qquad \left. \begin{aligned} a_{11}x_{12} + a_{12}x_{22} &= 0 \\ a_{21}x_{12} + a_{22}x_{22} &= 1 \end{aligned} \right\} \quad (2)$$

Solving systems (1) and (2) for x_{11}, x_{12}, x_{21}, and x_{22}, we have:

$$x_{11} = \frac{a_{22}}{a_{11}a_{22} - a_{12}a_{21}} \qquad x_{12} = \frac{-a_{12}}{a_{11}a_{22} - a_{12}a_{21}}$$

$$x_{21} = \frac{-a_{21}}{a_{11}a_{22} - a_{12}a_{21}} \qquad x_{22} = \frac{a_{11}}{a_{11}a_{22} - a_{12}a_{21}}$$

if and only if $(a_{11}a_{22} - a_{12}a_{21}) \neq 0$.

Let us define the *determinant* of A, written Det(A), as

$$\text{Det}(A) = a_{11}a_{22} - a_{12}a_{21}.$$

Therefore,

$$A^{-1} = \frac{1}{\text{Det}(A)} \begin{pmatrix} a_{22} & -a_{12} \\ -a_{21} & a_{11} \end{pmatrix}.$$

For example, if

$$A = \begin{pmatrix} 4 & 2 \\ -6 & 0 \end{pmatrix}, \text{ then}$$

$$A^{-1} = \frac{1}{12} \begin{pmatrix} 0 & -2 \\ 6 & 4 \end{pmatrix} = \begin{pmatrix} 0 & -\frac{1}{6} \\ \frac{1}{2} & \frac{1}{3} \end{pmatrix}.$$

7. The Solution of Linear Equations

We can use matrices to solve systems of linear equations. For example, consider the following system:

$$3U - 4V = -6$$

$$2U - 3V = -5. \qquad (3)$$

Using the definition of matrix multiplication, we can write system (3) in the following form:

$$\begin{pmatrix} 3 & -4 \\ 2 & -3 \end{pmatrix} \cdot \begin{pmatrix} U \\ V \end{pmatrix} = \begin{pmatrix} -6 \\ -5 \end{pmatrix}. \qquad (4)$$

The matrix equation (4) is of the pattern:

$$A \cdot X = B \qquad (5)$$

where A is the (2x2) matrix of the coefficients of U and V in system (3), X is the (2x1) matrix of variables U and V, and B is the (2x1) matrix of the constants.

Multiplying by A^{-1} in the matrix equation (5), we have:

$$A^{-1} \cdot A \cdot X = A^{-1} \cdot B.$$

But we know that:

$$A^{-1} \cdot A = I \text{ and } I \cdot X = X,$$

where I is the identity matrix for multiplication. Therefore we have:

$$X = A^{-1} \cdot B$$

as the solution of the matrix equation (5).

Let us use this method to find the solution of the matrix equation (4). We must find A^{-1}.

Now, $Det(A) = (3)(-3) - (2)(-4)$

$$= -9 + 8$$

$$= -1 .$$

Therefore, $A^{-1} = \dfrac{1}{-1} \cdot \begin{pmatrix} -3 & 4 \\ -2 & 3 \end{pmatrix} = \begin{pmatrix} 3 & -4 \\ 2 & -3 \end{pmatrix}.$

We then have:

$$\begin{pmatrix} U \\ V \end{pmatrix} = \begin{pmatrix} 3 & -4 \\ 2 & -3 \end{pmatrix}^{-1} \cdot \begin{pmatrix} -6 \\ -5 \end{pmatrix}$$

$$= \begin{pmatrix} 3 & -4 \\ 2 & -3 \end{pmatrix} \cdot \begin{pmatrix} -6 \\ -5 \end{pmatrix}$$

$$= \begin{pmatrix} 2 \\ 3 \end{pmatrix} .$$

Therefore, the solution of system (3) is $U = 2$ and $V = 3$. You can substitute these values in the equations of system (3) and thus check that $U = 2$ and $V = 3$ indeed satisfy the system.

The matrix procedure used to solve system (3) can be generalized to solve a system of n equations with n variables.

There are other ideas involving matrices which we shall not discuss in this book; you can delve more deeply into matrix theory by studying any number of published texts on the subject.

8-2 MATrix Statements in BASIC

1. List of MAT Statements

The purpose of our introducing a few ideas of matrices in Section 8-1 above is to utilize with some meaning a number of matrix operations in the BASIC language.

8-2 MATrix Statements in BASIC

When writing a matrix such as

$$A = \begin{pmatrix} a_{11} & a_{12} & a_{13} \\ a_{21} & a_{22} & a_{23} \end{pmatrix}$$

using the computer, we shall use the double subscript notation $A(I,J)$ to represent that entry in the I*th* row and the J*th* column. Thus, $A(1,1) = a_{11}$, $A(1,2) = a_{12}, \ldots, A(2,3) = a_{23}$.

We can define each entry in a matrix by means of LET statements, READ statements, or by the following double loop:

```
       .
       .
       .
100 INPUT R,C
110 FOR I=1 TO R
120     FOR J=1 TO C
130         INPUT A<I,J>
140     NEXT J
150 NEXT I
       .
       .
       .
```

We can also write programs to perform matrix operations such as finding the sum and product of matrices, finding the inverse and transpose of a matrix, etc. However, there are a number of matrix functions built into the BASIC language which will perform these operations for us. A list of these matrix statements, with a description of each, appears on the following page.

2. DIM, MAT READ, and MAT PRINT Statements

A DIM statement both reserves space for and determines the dimensions of a matrix. Step 10 of Example (1) below indicates that A is a 2 x 4 matrix and C is a 3 x 3 matrix. If no DIM statement is given for a matrix, the computer assumes that its dimensions are 10 x 10.

In Step 20 of Example (1), the numbers from the DATA (Steps 40 through 100) are read. The MAT READ statement enters numbers from the DATA list first for matrix A, then for matrix B, and finally for matrix C. Numbers are entered one row at a time for each matrix. That is, the computer first sets $A(1, 1) = 1$, $A(1, 2) = 2$, $A(1, 3) = 3$ and $A(1, 4) = 4$. Then, since the DIM statement in Step 10 indicates that A has 4 columns, the computer enters numbers into the second row of A: $A(2, 1) = 5$, $A(2, 2) = 6$, $A(2, 3) = 7$, and $A(2, 4) = 8$.

TABLE 4. MATrix Statements

Operation	Description
MAT READ A,B,C, . . .	READs the matrices A,B,C, . . . , where the DIMensions of each have been previously specified.
MAT PRINT A,B;C, . . .	PRINTs the matrices A,B,C . . . . The columns of A and C are printed in the five-column arrangement because of the comma which follows each. B is printed more compactly because of the semicolon.
MAT C = A+B	Adds the matrices A and B .
MAT C = A−B	Subtracts matrix B from A .
MAT C = (K)∗A	Multiplies matrix A by a scalar K, where K may be either a number or an expression. Note that K must be enclosed in parentheses.
MAT C = A∗B	Multiplies matrices A and B.
MAT B = TRN(A)	Transposes matrix A to A^T.
MAT B = INV(A)	Calculates the inverse of matrix A.
MAT A = ZER	Sets each entry of A to zero.
MAT A = CON	Sets each entry of A to one.
MAT A = IDN	Sets up an identity matrix A.
MAT INPUT A	Inputs a vector A.

Since the DIM statement of Step 10 indicates that A has 2 rows, the computer begins to enter values for matrix B. The values for matrix B are read one row at a time, beginning with the next item in the DATA. That is, B(1, 1) = 99, B(1, 2) = 10, . . . , B(1, 10) = 18. Since B

EXAMPLE (1)

PROGRAM:

```
10 DIM A(2,4),C(3,3)
20 MAT READ A,B,C
30 MAT PRINT A,B;C
40 DATA 1,2,3,4,5,6,7,8,99,10,11,12,13,14,15,16,17,18,19,20,21,22
50 DATA 23,24,25,26,27,28,29,30,31,32,33,34,35,36,37,38,39,40,41
60 DATA 42,43,44,45,46,47,48,49,50,51,52,53,54,55,56,57,58,59,60
70 DATA 61,62,63,64,65,66,67,68,69,70,71,72,73,74,75,76,77,78,79
80 DATA 80,81,82,83,84,85,86,87,88,89,90,91,92,93,94,95,96,97,98
90 DATA 99,98,97,96,95,94,93,92,91,90,109,110,111,112,113,114,115
100 DATA 116,117
110 END
```

OUTPUT:

1	2	3	4
5	6	7	8

99	10	11	12	13	14	15	16	17	18
19	20	21	22	23	24	25	26	27	28
29	30	31	32	33	34	35	36	37	38
39	40	41	42	43	44	45	46	47	48
49	50	51	52	53	54	55	56	57	58
59	60	61	62	63	64	65	66	67	68
69	70	71	72	73	74	75	76	77	78
79	80	81	82	83	84	85	86	87	88
89	90	91	92	93	94	95	96	97	98
99	98	97	96	95	94	93	92	91	90

109	110	111
112	113	114
115	116	117

is not dimensioned in a DIM statement, the computer assumes it to be a 10 x 10 matrix. The computer thus reads the next 10 items of the DATA for the second row: B(2, 1) = 19, B(2, 2) = 20, . . . , B(2, 10) = 28. This process continues until B(10, 10) is set equal to 90.

After the computer reads the DATA for B, it then assigns the remaining items of the DATA to matrix C in the same fashion. That is, C(1, 1) = 109, C(1, 2) = 110, C(1, 3) = 111, C(2, 1) = 112, C(2, 2) = 113, ..., C(3, 3) = 117.

The computer then executes Step 30 by printing the matrices A, B, and C, as you can see in the OUTPUT. Notice that the MAT PRINT statement prints matrices A and C in the five-column arrangement because of the comma and lack of punctuation respectively, whereas B is printed in a more compact form because of the semicolon. Note in the OUTPUT that two spaces separate each entry in matrix B: one space for the sign and one space following each entry. If an entry had a negative value, then the preceding space would contain a negative sign (-).

The MAT READ statement may also be used to determine the dimension of a matrix during the execution of a program, as well as to read the entries of the matrix, as you can see in Example (2) below. Step 10 reads H and L. Step 20 defines E as an H x L (2 x 3) matrix and reads each entry of E from the DATA of Step 50. Step 30 prints matrix E in compact form. Note that the DATA in Steps 40 and 50 could have been written on one line. We used two lines for easier reading. Also note that a DIM statement would have been required if the DATA of Step 40 called for a matrix with dimensions greater than 10 x 10.

EXAMPLE (2)

PROGRAM:

```
10 READ H,L
20 MAT READ E(H,L)
30 MAT PRINT E;
40 DATA 2,3
50 DATA 1,2,3,4,5,6
60 END
```

OUTPUT:

```
 1  2  3
 4  5  6
```

---

3. The Zero Matrix, Identity Matrix, and Matrix of Ones

Example (3) below illustrates the following three MAT statements:

$$\text{MAT } Z = \text{ZER } (R, C)$$

$$\text{MAT } D = \text{CON } (J, K)$$

$$\text{MAT } I = \text{IDN } (N, N).$$

The above three MAT statements can also be used to determine the dimensions of a matrix during the execution of a program. In Example (3), Step 20 "fills out" an R x C matrix Z with zeros. Since R = 5 and C = 6 from the READ and DATA statements of Steps 10 and 130, Z is a 5 x 6 matrix of zeros.

Step 30 produces a J x K matrix D of ones. Since J = 4 and K = 8 from the READ and DATA statements, Step 30 defines a 4 x 8 matrix of ones.

Step 40 constructs a square N x N identity matrix I. Recall that the identity matrix contains 1's on the main diagonal, and 0's in all other entries. Since N = 6 from the READ-DATA statements, a 6 x 6 identity matrix I is stored.

Steps 50, 80, and 110 print an appropriate heading for each matrix, and Steps 60, 90, and 120 print the corresponding matrices in compact format, as you can see in the OUTPUT.

EXAMPLE (3)

PROGRAM:

```
10 READ R,C,J,K,N
20 MAT Z=ZER(R,C)
30 MAT D=CON(J,K)
40 MAT I=IDN(N,N)
50 PRINT "ZERO MATRIX: DIMENSIONS"R;"X"C
60 MAT PRINT Z;
70 PRINT
80 PRINT "MATRIX OF ONES: DIMENSIONS"J;"X"K
90 MAT PRINT D;
100 PRINT
110 PRINT "IDENTITY MATRIX: DIMENSIONS"N;"X"N
120 MAT PRINT I;
130 DATA 5,6,4,8,6
140 END
```

OUTPUT:

```
ZERO MATRIX: DIMENSIONS 5 X 6

 0  0  0  0  0  0
 0  0  0  0  0  0
 0  0  0  0  0  0
 0  0  0  0  0  0
 0  0  0  0  0  0

MATRIX OF ONES: DIMENSIONS 4 X 8

 1  1  1  1  1  1  1  1
 1  1  1  1  1  1  1  1
 1  1  1  1  1  1  1  1
 1  1  1  1  1  1  1  1

IDENTITY MATRIX: DIMENSIONS 6 X 6

 1  0  0  0  0  0
 0  1  0  0  0  0
 0  0  1  0  0  0
 0  0  0  1  0  0
 0  0  0  0  1  0
 0  0  0  0  0  1
```

As with the MAT READ statement, the three statements above (ZER, CON, IDN) may be used without subscripts. In Example (4), we first define Z as a 3 x 3 zero matrix in Step 10. Then in Step 40 we redefine Z as an identity matrix. Note from the OUTPUT of Step 60 that the dimensions of Z are still 3 x 3, as defined in Step 10.

Recall from Paragraph 2 above that, on most systems, the computer DIMensions a matrix before executing the program, regardless of where the DIM statement for that matrix is placed in the program. However, a matrix can be redimensioned by any one of the four MAT statements of READ, ZER, CON, or IDN.

Look at Example (5). The DIM statement first DIMensions F as a

EXAMPLE (4)

PROGRAM:

```
10 MAT Z=ZER(3,3)
20 PRINT "3X3 ZERO MATRIX:"
30 MAT PRINT Z;
40 MAT Z=IDN
50 PRINT "3X3 IDENTITY MATRIX:"
60 MAT PRINT Z;
70 END
```

OUTPUT:

```
3X3 ZERO MATRIX:

 0   0   0

 0   0   0

 0   0   0

3X3 IDENTITY MATRIX:

 1   0   0

 0   1   0

 0   0   1
```

12 x 15 matrix and reserves space for it. Then the MAT READ statement of Step 20 redimensions F as a 3 x 4 matrix and reads the entries of F from the DATA. However, the computer still reserves space for a 12 x 15 matrix. The statement MAT F = ZER in Step 30 replaces each of the entries of F with zeros. The OUTPUT indicates that F is indeed a 3 x 4 matrix, since it was last given the dimensions 3 x 4 in the MAT READ statement of Step 20.

EXAMPLE (5)

PROGRAM:

```
10 DIM F(12,15)
20 MAT READ F(3,4)
30 MAT F=ZER
40 MAT PRINT F;
50 DATA 1,2,3,4, 4,3,2,1, 5,6,7,8
60 END
```

OUTPUT:

```
0    0    0    0
0    0    0    0
0    0    0    0
```

4. The Sum, The Differences of Matrices, and Scalar Product Statements

Example (6) below illustrates the use of the MAT sum, difference, and scalar product statements. Notice that the dimensions of matrices A (sum), S (difference) and M (scalar product) are not given in the program; the computer automatically determines them. However, the dimensions of B and C must be included in the program, as you can see in Step 10. Note that the dimensions of B and C must be the same, since matrix sum and difference are defined only for matrices of the same dimensions. The computer would print an appropriate ERROR message if the dimensions of B and C were not the same.

Also note that the scalar F in Step 150 must be enclosed in parentheses.

EXAMPLE (6)

PROGRAM:

```
10 MAT READ B(2,3),C(2,3)
20 REM---------SUM----------
30 MAT A=B+C
40 PRINT "THE SUM IS:"
50 MAT PRINT A;
60 PRINT
70 REM---------DIFFERENCE----------
80 MAT S=B-C
90 PRINT "THE DIFFERENCE IS:"
100 MAT PRINT S;
110 PRINT
120 REM---------SCALAR PRODUCT----------
130 READ F
140 PRINT "MULTIPLYING THE FIRST MATRIX BY"F;"GIVES:"
150 MAT M=(F)*B
160 MAT PRINT M
170 DATA 2,3,4, 5,6,7, -1,2,5, 0,-2,4, 0.5
180 END
```

OUTPUT:

THE SUM IS:

```
1    5    9
5    4   11
```

THE DIFFERENCE IS:

```
3    1   -1
5    8    3
```

MULTIPLYING THE FIRST MATRIX BY 0.5 GIVES:

```
1          1.5         2
2.5        3           3.5
```

5. The Transpose, Product, Inverse, and Determinant Statements

Example (7) below illustrates the following MAT statements:

$$\begin{aligned} &\text{MAT C} = \text{TRN(B)} \\ &\text{MAT P} = \text{A}*\text{B} \\ &\text{MAT I} = \text{INV(A)} \\ &\text{DET .} \end{aligned}$$

Steps 30 through 50 determine and print the transpose of matrix B; Steps 80 through 100 calculate and print the product of A and B; Steps 130 through 150 find and print the inverse of A. Notice that the computer automatically determines the dimensions of C, P, and I.

To calculate the determinant of a matrix A, the computer must first find the inverse of A, as in Step 130. At a given point in a program, the function DET is set equal to the determinant of that matrix whose inverse was last calculated in the program. Thus in Step 180 of Example (7), DET is set equal to the determinant of A, since the last inverse the computer calculated was INV(A).

Notice in Step 10 that the dimensions of A and B are 3 x 3 and 3 x 4 respectively. Hence the OUTPUT of the product matrix P in Step 100 shows that the dimensions of P are 3 x 4, as expected. If the

EXAMPLE (7)

PROGRAM:

```
10 MAT READ A(3,3),B(3,4)
20 REM------------TRANSPOSE------------
30 MAT C=TRN(B)
40 PRINT "THE TRANSPOSE OF MATRIX B IS:"
50 MAT PRINT C;
60 PRINT
70 REM------------MATRIX PRODUCT------------
80 MAT P=A*B
90 PRINT "THE MATRIX PRODUCT IS:"
100 MAT PRINT P
110 PRINT
120 REM------------INVERSE------------
130 MAT I=INV(A)
140 PRINT "THE INVERSE OF MATRIX A IS:"
150 MAT PRINT I
160 PRINT
170 REM------------DETERMINANT------------
180 PRINT "THE DETERMINANT OF MATRIX A IS"DET
190 PRINT
200 REM------------PRODUCT OF MATRIX AND INVERSE------------
210 MAT R=A*I
220 PRINT "PRODUCT OF MATRIX A AND ITS INVERSE:"
230 MAT PRINT R
240 DATA 1,1,1, 3,4,5, 1,2,1
250 DATA 2,1,0,0, 0,0,1,1, 1,0,0,1
260 END
```

OUTPUT:

THE TRANSPOSE OF MATRIX B IS:

```
2   0   1

1   0   0

0   1   0

0   1   1
```

THE MATRIX PRODUCT IS:

```
3           1           1           2

11          3           4           9

3           1           2           3
```

```
THE INVERSE OF MATRIX A IS:

 3              -0.5           -0.5
-1.              0              1.
-1.              0.5           -0.5

THE DETERMINANT OF MATRIX A IS 2
PRODUCT OF MATRIX A AND ITS INVERSE:

 1.              0             -3.72529 E-9
 0               1              0
 1.49012 E-8    0              1
```

product of two matrices is to be calculated, but the dimensions of those matrices are not properly given according to the definition of matrix product, then the computer will print an appropriate ERROR message.

From Steps 210 through 230 we would expect the product of a matrix and its inverse to be the identity matrix. In the OUTPUT of Step 230, the matrix which is printed is not exactly the identity matrix because of the round-off error of the computer.

6. Matrix Equality

We may set one matrix equal to another matrix by a statement of the form

$$\text{MAT } D = C \;,$$

as illustrated in Example (8). Step 10 reads a 2 x 4 matrix C. In Step 20 matrix D is defined to be equal to C. The OUTPUT of Steps 40 and 60 show that indeed D has the same entries as C. Note that D need not be dimensioned in the program; it is automatically given the same dimensions as C.

On some systems, the statement MAT D = C is not allowed. However, the same result may be achieved with the statement MAT D = (1) * C.

EXAMPLE (8)

PROGRAM:

```
10 MAT READ C(2,4)
20 MAT D=C
30 PRINT "MATRIX C:"
40 MAT PRINT C;
50 PRINT "MATRIX D:"
60 MAT PRINT D;
70 DATA 1,3,5,7, 2,4,6,8
80 END
```

OUTPUT:

MATRIX C:

```
1   3   5   7
2   4   6   8
```

MATRIX D:

```
1   3   5   7
2   4   6   8
```

7. The Same Matrix Variable on Both Sides of an Equal Sign

The same matrix variable may appear on both sides of the equal sign in the following four MAT statements:

$$\begin{aligned} &\text{MAT } A = A + B \\ &\text{MAT } A = A - B \\ &\text{MAT } A = (K)*A, \text{ where K is a scalar} \\ &\text{MAT } A = \text{INV}(A) \ . \end{aligned}$$

In Step 30 of Example (9), the computer calculates the matrix sum of B and C and places the resulting matrix back in storage area B. Step 80 finds the difference between E and C and stores the result in matrix E. In Step 140, the entries of C are multiplied by 3*2 or 6 and placed back in C.

EXAMPLE (9)

PROGRAM:

```
10 MAT READ B(2,3),C(2,3),E(2,3)
20 REM---------SUM----------
30 MAT B=B+C
40 PRINT "THE SUM IS:"
50 MAT PRINT B
60 PRINT
70 REM---------DIFFERENCE----------
80 MAT E=E-C
90 PRINT "THE DIFFERENCE IS:"
100 MAT PRINT E
110 PRINT
120 REM---------SCALAR PRODUCT----------
130 PRINT "MULTIPLYING MATRIX C BY"3*2;"GIVES:"
140 MAT C=(3*2)*C
150 MAT PRINT C
160 DATA 2,3,4, 4,6,7, -1,2,5, 0,-2,4, 3,2,7, 1,0,6
170 END
```

OUTPUT:

THE SUM IS:

1	5	9
4	4	11

THE DIFFERENCE IS:

4	0	2
1	2	2

MULTIPLYING MATRIX C BY 6 GIVES:

-6	12	30
0	-12	24

In Step 30 of Example (10), the inverse of matrix B is calculated and placed back in storage area B. Since the OUTPUT is the identity matrix, we know that the statement in Step 30 sets B equal to its own inverse. (Statements such as MAT B = INV(B) are not allowed on some systems.)

EXAMPLE (10)

PROGRAM:

```
10 MAT READ B(2,2)
20 MAT C=B
30 MAT B=INV(B)
40 MAT D=C*B
50 PRINT "PRODUCT OF A MATRIX AND ITS INVERSE:"
60 MAT PRINT D;
70 DATA 4,3,2,1
80 END
```

OUTPUT:

PRODUCT OF A MATRIX AND ITS INVERSE:

1 0

0 1

8. Illegal MAT Statements

There are some operations with MAT statements which the computer does not accept. When this occurs, the computer prints an appropriate ERROR message, as you can see in Example (11) below.

The ERROR message of Step 30 indicates that we cannot have more than one matrix operator in one MAT statement. We must use one MAT statement for each matrix operator in a program. The ERROR messages for Steps 40 and 50 indicate that we cannot have the same variables on both sides of the equal sign in MAT transpose and MAT multiplication statements.

9. More About DIM Statements; Vectors

As with variables having single subscripts, the statement DIM A(R, C) reserves space for subscripted variables from A (0, 0) to A (R, C), or space for $(R+1)*(C+1)$ variables. However, although the computer reserves space for this number of variables, the MAT statements on most systems interpret the matrix A as an R x C dimensioned matrix whose entries are the variables A (I, J), where I = 1, 2, 3, ..., R and J = 1, 2, 3, ..., C. That is, the first row of the matrix has the subscript I = 1, and the first column has the subscript J = 1.

The reason for this apparent discrepancy is for our convenience. When we speak of an entry in the Ith row and the Jth column of a matrix A, we usually mean the element A (I, J). If the computer read the matrix whose entries are A (I, J) for I = 0, 1, 2, ..., R-1 and J =

EXAMPLE (11)

PROGRAM:

```
10 REM-----------ILLEGAL STATEMENTS-----------
20 MAT READ A(2,2), B(2,2), C(2,2)
30 MAT D=A+B-C
40 MAT A=TRN(A)
50 MAT B=B*C
60 MAT PRINT A, B, D
70 DATA 1,2,3,4, 4,3,2,1, 0,9,8,7
80 END
```

OUTPUT:

```
ILLEGAL FORMAT IN 30
ILLEGAL MAT TRANSPOSE IN 40
ILLEGAL MAT MULTIPLE IN 50
```

0, 1, 2, ..., C-1, then we would need to remind ourselves constantly that the entry a_{ij} is stored in A (I-1, J-1).

However, we may also store and operate on a matrix of the form A(0, J) or B (I, 0), in which we utilize the zero row or the zero column of a matrix. We may think of these zero row or zero column matrices as *vectors*. (The computer also interprets the column vector B(I) as B (I, 0) and stores this vector in the zero column.)

Look at Example (12) below. Step 10 dimensions a four-component row vector A and a seven-component column vector B. Step 20 reads the entries of A and B. Step 30 prints A as a *column* vector because of the lack of punctuation. The semicolon in Step 40 directs the computer to print B as a *row* vector in compact format. (A comma after B in Step 40 would have caused the vector B to be printed in the five-column arrangement; the last two entries would have been printed on the next line.)

EXAMPLE (12)

PROGRAM:

```
10 DIM A(0,4), B(7)
20 MAT READ A, B
30 MAT PRINT A
40 MAT PRINT B;
50 DATA 1,2,3,4, 1,2,3,4,5,6,7
60 END
```

OUTPUT:

1
2
3
4

1 2 3 4 5 6 7

When using a matrix of dimensions significantly less than 10 x 10, the use of a DIM statement is recommended. For example, if you were using a matrix B of dimensions 4 x 6, the statement DIM B (4, 6) would instruct the computer to reserve space for 5∗7 = 35 variables, instead of the 11∗11 = 121 variables which it automatically reserves if no DIM statement is included. Reserving only the necessary space for matrices might be important in long programs, or in programs which handle other matrices with large dimensions.

As stated above, although the MAT statements do not normally operate on the zero row and the zero column, MAT statements may affect the values of variables which are specifically stored in this row and column. These values are affected because of the redimensioning of matrices which is sometimes required in MAT statements. For example, if we set A (1, 0) = 2 in a LET statement, and then give the instruction MAT READ A(2, 2), the value of A(1, 0) will be automatically reset to 0. Thus it is best not to use the zero row and zero column when using MAT statements in a program, unless you specifically need them as vectors.

10. N-Dimensional Arrays

Although BASIC does not include commands which can operate on n-dimensional arrays, we can write a program which accomplishes the same objective, as in Examples (13) and (14) below.

Example (13) illustrates how we can assign one number to an ordered triplet (I, J, K) by means of the DEF statement in Step 20. The computer reads the DATA of Step 110 in the MAT READ B(3) statement of Step 30 and thus assigns 2 to B(1), 3 to B(2), and 2 to B(3). A triple-nested loop is then executed in Steps 40 through 100. On the first time through the triple loop, I =1, J =1, and K = 1 from Steps 40, 50, and 60 respectively. Then the computer: (1) reads the FNC (I, J, K) statement in Step 70; (2) searches through the program and locates the DEF FNC statement in Step 20; (3) evaluates FNC (1, 1, 1) = (1-1)*3*2 + (1-1)*2 + 1 = 1; (4) and then prints the values of I, J, K and FNC (I, J, K) in Step 70. K is then set equal to 2 in Step 80, FNC (1, 1, 2) is evaluated, and the values of I, J, K and FNC (I, J, K) are again printed. The triple-nested loop continues as J is set to 2 and K to 1, FNC (1, 2, 1) is evaluated, and the third line of the OUTPUT is printed. The computer continues to execute the triple-nested loop until the twelfth line of the OUTPUT is printed.

We have thus associated with each ordered triplet (I, J, K) a unique number FNC (I, J, K). If I, J and K are respectively less than or equal to B(1), B(2) and B(3), exactly one combination of (I, J, K) will produce a given value of FNC. This value of FNC will be a number between 1 and B(1) * B(2) * B(3), the number of components in the array.

EXAMPLE (13)

PROGRAM:

```
10 PRINT " I"," J"," K","FNC(I,J,K)"
20 DEF FNC(I,J,K)=(I-1)*B(2)*B(3)+(J-1)*B(3)+K
30 MAT READ B(3)
40 FOR I=1 TO B(1)
50    FOR J=1 TO B(2)
60       FOR K=1 TO B(3)
70          PRINT I,J,K,FNC(I,J,K)
80       NEXT K
90    NEXT J
100 NEXT I
110 DATA 2,3,2
120 END
```

OUTPUT:

I	J	K	FNC(I,J,K)
1	1	1	1
1	1	2	2
1	2	1	3
1	2	2	4
1	3	1	5
1	3	2	6
2	1	1	7
2	1	2	8
2	2	1	9
2	2	2	10
2	3	1	11
2	3	2	12

Now that we have associated a single value FNC (I, J, K) with each set of coordinates I, J, K, we may use the vector A (FNC (I, J, K)) to store the array, as shown in Example (14). For our purposes, the vector A (FNC (I, J, K)) behaves exactly like a variable of the form A (I, J, K). Thus we have simulated in BASIC an ordered triplet by using this technique.

Step 100 of Example (14) reserves enough space for an array of 100 components. Step 120 reads the dimensions of the array from the DATA in Step 340. Step 130 prints the number of components in the array. The loops in Steps 160 through 230 find the values of the components of the array and store them in the vector A. The first time through the loops, Step 190 assigns to L the value of the next item in the DATA list (53). Since I = J = K = 1, FNC (I, J, K) = FNC (1, 1, 1) = 0∗4∗5+0∗5+1 = 1. Also, I+J+K+L =1+1+1+53 = 56. Therefore, 56 is assigned to the variable A(1). Similarly, the second time through the loop, I = J = 1, K = 2 and L = 14. Thus, since FNC (1, 1, 2) = 0∗4∗5+0∗5+2 = 2 and I+J+K+L = 1+1+2+14 = 18, A (2) is set equal to 18. This process continues until FNC (2, 4, 5) = 1∗4∗5+3∗5+5 = 40 and I+J+K+L = 2+4+5+84 = 95; hence A(40) is set equal to 95.

Now that the computer has calculated and stored each of the entries in our array, we would like to be able to reference these entries using three variables, as if the triple-subscript notation A (I, J, K) existed on the computer. We can easily simulate an ordered triplet, as the OUTPUT of Steps 250 to 320 illustrates.

The computer asked us from Step 260 WHAT ARE THE COORDINATES OF THE COMPONENT in which we are interested. We first typed 1, 1 and 1 for X, Y, and Z. Since FNC (1, 1, 1) = 0∗4∗5+0∗5+1 = 1, then the computer printed the number stored in A(1), namely, 56. The computer again asked us for the coordinates of the entry in which we are interested. We typed 2, 4, and 5 for X, Y, and Z. Since FNC (2, 4, 5) = 1∗4∗5+3∗5+5 = 40, the computer printed the fortieth component of the vector A, namely 95. We can thus determine the number stored in any component of vector A, and in this fashion we can simulate the ordered triplet A(I, J, K). Note that Step 280 allows us to stop the program by typing 0 for any one of the three coordinates X, Y, and Z.

If we wish to check for error conditions, we can use a multiline DEFinition for FNC to ensure that the arguments (or subscripts) I, J, and K are respectively less than B(1), B(2), and B(3), greater than 1, and integral.

Since functions in BASIC can have any number of arguments, we can easily modify Example (14) to simulate an n-dimensional array.

EXAMPLE (14)

PROGRAM:

```
100 DIM A(100)
110 DEF FNC(I,J,K)=(I-1)*B(2)*B(3)+(J-1)*B(3)+K
120 MAT READ B(3)        '----------DIMENSIONS OF ARRAY----------
130 PRINT "YOUR ARRAY CONTAINS"B(1)*B(2)*B(3);"COMPONENTS."
140 PRINT
150 REM----------READING COMPONENTS OF ARRAY----------
160 FOR I=1 TO B(1)
170     FOR J=1 TO B(2)
180         FOR K=1 TO B(3)
190             READ L
200             A(FNC(I,J,K))=I+J+K+L
210         NEXT K
220     NEXT J
230 NEXT I
240 REM----------ENTERING COORDINATES----------
250 PRINT
260 PRINT "WHAT ARE THE COORDINATES OF THE COMPONENT";
270 INPUT X,Y,Z
280 IF X*Y*Z=0 THEN 370
290 REM-----TYPING COMPONENTS WITH SPECIFIED COORDINATES-----
300 PRINT "THE NUMBER IS"A(FNC(X,Y,Z))
310 PRINT
320 GOTO 250
330 REM----------DIMENSIONS AND VALUES OF L----------
340 DATA 2,4,5
350 DATA 53,14,27,92,31,39,46,57,17,8,3,6,16,21,82,77,1,35,71,65
360 DATA 11,61,25,2,66,94,10,45,6,30,51,2,19,26,78,42,7,52,56,84
370 END
```

OUTPUT:

```
YOUR ARRAY CONTAINS 40 COMPONENTS.


WHAT ARE THE COORDINATES OF THE COMPONENT? 1,1,1
THE NUMBER IS 56


WHAT ARE THE COORDINATES OF THE COMPONENT? 2,4,5
THE NUMBER IS 95


WHAT ARE THE COORDINATES OF THE COMPONENT? 1,4,5
THE NUMBER IS 75


WHAT ARE THE COORDINATES OF THE COMPONENT? 1,2,3
THE NUMBER IS 63


WHAT ARE THE COORDINATES OF THE COMPONENT? 0,1,1
```

11. MAT Statements with Strings

Strings can also be manipulated in a fashion similar to manipulating numbers using the commands MAT READ, MAT PRINT, and MAT INPUT with string variables, as you can see in Example (15) below.

EXAMPLE (15)

PROGRAM:

```
10 DIM B$(4)
20 MAT READ A$(3),B$
30 MAT PRINT A$
40 MAT PRINT B$;
50 PRINT "WHAT ARE YOUR STRINGS";
60 MAT INPUT C$
70 PRINT "YOU HAVE TYPED"NUM;"STRINGS. THEY ARE:"
80 MAT PRINT C$,
90 DATA ABC,DEF,GHI,JKL,MNO,PQR,STU
100 END
```

OUTPUT:

```
ABC
DEF
GHI


JKLMNOPQRSTU

WHAT ARE YOUR STRINGS? SUM,PRODUCT,QUOTIENT,DIFFERENCE
YOU HAVE TYPED 4 STRINGS. THEY ARE:

SUM            PRODUCT        QUOTIENT       DIFFERENCE
```

Notice that we may specify the number of strings either in a DIM statement as in Step 10, or in a MAT READ statement as in Step 20. However, the MAT INPUT C$ statement in Step 60 does not require a DIMension statement for C$ if there are less than ten strings. NUM is set equal to the number of components entered for the vector C$. The MAT INPUT statement for strings thus behaves in exactly the same manner as it does for numeric input.

Step 20 illustrates how to read a series of strings using the MAT READ statement, and Steps 30, 40, and 80 show how to print the strings by means of the MAT PRINT command. Notice in the OUTPUT that the MAT PRINT A$ statement of Step 30 caused the computer to type the strings contained in A$(1), A$(2), and A$(3) vertically, because there is no punctuation mark to the right of A$ in Step 30. However, the semicolon of Step 40 caused the computer to print the contents of B$ horizontally, with no spaces between the strings. The comma at the end of Step 80 produced an OUTPUT of C$ in the five-column arrangement.

EXERCISES 8-2

1. Write a program which shows, using several test cases, the following properties of matrix addition. Let A, B, and C be (n x n) matrices, Z the zero matrix, and p and q scalars.
    a. A+Z = Z+A = A
    b. A+B = B+A
    c. A+(-A) = Z
    d. A+(B+C) = (A+B)+C
    e. p(A+B) = pA+pB
    f. (p+q)A = pA+qA
    g. p(qA) = (pq)A

2. Write a program which shows, using a number of test cases, that:

$$A \cdot I = I \cdot A = A \quad \text{and}$$

$$A \cdot A^{-1} = A^{-1} \cdot A = I,$$

where A is an (n x n) matrix, A^{-1} is the inverse of A, and I is the identity matrix of multiplication of dimensions (n x n).

3. Write a program which shows, using several test cases, that matrix multiplication is not commutative, i.e.,

$$A \cdot B \neq B \cdot A,$$

where A and B are two (n x n) matrices, neither of which is the identity matrix of multiplication, and $A \neq B^{-1}$.

4. Write a program which shows, using a number of test cases, that

$$(A^T)^T = A \quad \text{and}$$

$$(A \cdot B)^T = B^T \cdot A^T ,$$

where A is an (n x n) matrix, B is an (n x n) matrix, A^T is the transpose of A, $(A^T)^T$ is the transpose of the transpose of A, and $(A \cdot B)^T$ is the transpose of the product of A and B.

5. Write a program which shows, using a number of test cases, that

$$(A^{-1})^{-1} = A \quad \text{and}$$

$$(A \cdot B)^{-1} = B^{-1} \cdot A^{-1} ,$$

where A is an (n x n) matrix whose $Det(A) \neq 0$, and B is an (n x n) matrix whose $Det(B) \neq 0$, and $(A^{-1})^{-1}$ is the inverse of the inverse of A.

8-3. The Solution of Linear Systems Using Matrices

1. The Matrix Solution of n Equations with n Variables

Review Paragraph 7 of Section 8-1. Recall that we may write a system of two equations with two variables such as

$$3U - 4V = -6$$

$$2U - 3V = -5 \tag{1}$$

by the matrix equation

$$\begin{pmatrix} 3 & -4 \\ 2 & -3 \end{pmatrix} \cdot \begin{pmatrix} U \\ V \end{pmatrix} = \begin{pmatrix} -6 \\ -5 \end{pmatrix}. \tag{2}$$

We may write the solution of Equation (2) as follows:

$$\begin{pmatrix} U \\ V \end{pmatrix} = \begin{pmatrix} 3 & -4 \\ 2 & -3 \end{pmatrix}^{-1} \cdot \begin{pmatrix} -6 \\ -5 \end{pmatrix}.$$

Let us generalize the above example to solve n linear equations with n variables each by means of matrices. We write the system of equations as follows:

8-3 The Solution of Linear Systems Using Matrices

$$a_{11}x_1 + a_{12}x_2 + a_{13}x_3 + \cdots + a_{1n}x_n = b_1$$
$$a_{21}x_1 + a_{22}x_2 + a_{23}x_3 + \cdots + a_{2n}x_n = b_2$$
$$a_{31}x_1 + a_{32}x_2 + a_{33}x_3 + \cdots + a_{3n}x_n = b_3 .$$
$$\vdots$$
$$a_{n1}x_1 + a_{n2}x_2 + a_{n3}x_3 + \cdots + a_{nn}x_n = b_n \quad (3)$$

We may write System (3) as the following matrix equation:

$$\begin{pmatrix} a_{11} & a_{12} & a_{13} & \cdots & a_{1n} \\ a_{21} & a_{22} & a_{23} & \cdots & a_{2n} \\ a_{31} & a_{32} & a_{33} & \cdots & a_{3n} \\ \vdots \\ a_{n1} & a_{n2} & a_{n3} & \cdots & a_{nn} \end{pmatrix} \cdot \begin{pmatrix} x_1 \\ x_2 \\ x_3 \\ \vdots \\ x_n \end{pmatrix} = \begin{pmatrix} b_1 \\ b_2 \\ b_3 \\ \vdots \\ b_n \end{pmatrix} \quad (4)$$

The matrix solution of Equation (4) is as follows:

$$\begin{pmatrix} x_1 \\ x_2 \\ x_3 \\ \vdots \\ x_n \end{pmatrix} = \begin{pmatrix} a_{11} & a_{12} & a_{13} & \cdots & a_{1n} \\ a_{21} & a_{22} & a_{23} & \cdots & a_{2n} \\ a_{31} & a_{32} & a_{33} & \cdots & a_{3n} \\ \vdots \\ a_{n1} & a_{n2} & a_{n3} & \cdots & a_{nn} \end{pmatrix}^{-1} \cdot \begin{pmatrix} b_1 \\ b_2 \\ b_3 \\ \vdots \\ b_n \end{pmatrix} \quad (5)$$

When we write the program to solve a system of n linear equations with n variables each, we realize from Equation (5) above that all that is necessary for the computer to find is the inverse of the n x n matrix of coefficients, and then to find the product of that matrix and the n x 1 matrix of constants.

2. Program (14): The Solution of Linear Systems Using Matrices

The following program uses the procedure given above to solve a system of n simultaneous linear equations with n variables, using the MAT statements on the computer.

PROGRAM (14):

```
100 DIM A(30,30),B(30),X(30)
110 PRINT "THIS PROGRAM SOLVES A SYSTEM OF SIMULTANEOUS LINEAR ";
120 PRINT "EQUATIONS."
130 PRINT
140 REM----------ENTERING COEFFICIENTS----------
150 PRINT "HOW MANY VARIABLES ARE THERE IN YOUR SYSTEM OF EQUATIONS";
160 INPUT U
170 PRINT
180 PRINT
190 MAT A=ZER(U,U)
200 FOR R=1 TO U
210     PRINT "COEFFICIENTS OF ROW"R;
220     MAT INPUT B
230     FOR I=1 TO U
240        A(R,I)=B(I)
250     NEXT I
260     PRINT
270 NEXT R
280 PRINT
290 REM----------INVERTING MATRIX OF COEFFICIENTS----------
300 MAT A=INV(A)
310 IF ABS(DET)>=1E-7 THEN 350
320 PRINT "THERE IS NO UNIQUE SOLUTION TO YOUR SYSTEM OF EQUATIONS."
330 GOTO 530
340 REM----------ENTERING CONSTANTS----------
350 MAT B=ZER(U)
360 PRINT "WHAT ARE YOUR CONSTANTS";
370 MAT INPUT B
380 PRINT
390 IF NUM=U THEN 430
400 PRINT "THE NUMBER OF CONSTANTS SHOULD EQUAL THE NUMBER OF ";
410 PRINT "VARIABLES."
420 GOTO 350
430 PRINT
440 PRINT
450 REM----------FINDING AND PRINTING SOLUTION----------
460 MAT X=A*B
470 PRINT "THE SOLUTION IS:"
480 PRINT
490 FOR H=1 TO U
500     PRINT "X("H;")="X(H)
510 NEXT H
520 REM----------REPEAT?----------
530 PRINT
540 PRINT
550 PRINT
560 PRINT "DO YOU WISH TO RUN AGAIN";
570 INPUT Q$
580 PRINT
590 IF Q$="Y" THEN 130
600 END
```

OUTPUT:

THIS PROGRAM SOLVES A SYSTEM OF SIMULTANEOUS LINEAR EQUATIONS.
HOW MANY VARIABLES ARE THERE IN YOUR SYSTEM OF EQUATIONS ?2

COEFFICIENTS OF ROW 1 ?3,-4
COEFFICIENTS OF ROW 2 ?2,-3

WHAT ARE THE CONSTANTS ?-6,-5

THE SOLUTION IS:
X(1)= 2
X(2)= 3

DO YOU WISH TO RUN AGAIN ?YES

HOW MANY VARIABLES ARE THERE IN YOUR SYSTEM OF EQUATIONS ?3

COEFFICIENTS OF ROW 1 ?1,-2,-3
COEFFICIENTS OF ROW 2 ?1,1,-1
COEFFICIENTS OF ROW 3 ?2,-3,-5

WHAT ARE THE CONSTANTS ?3,2,5

THE SOLUTION IS:
X(1)=-1.
X(2)= 1.
X(3)=-2.

DO YOU WISH TO RUN AGAIN ?YES

HOW MANY VARIABLES ARE THERE IN YOUR SYSTEM OF EQUATIONS ?5

COEFFICIENTS OF ROW 1 ?1.42,   3.651, 0,   -6,   55.6
COEFFICIENTS OF ROW 2 ?-5.76, -2.1,   32.4,   667,   3
COEFFICIENTS OF ROW 3 ?-1,   0,   0,   2.361,   3.75
COEFFICIENTS OF ROW 4 ?4.721, -3.52,   7.5439, 1,   4.37
COEFFICIENTS OF ROW 5 ?-5.391,-6.37,   4.98,   2.75,   1.504

```
WHAT ARE THE CONSTANTS ?5.96,  1.42,   3.61,   3.99,   4.682


THE SOLUTION IS:

X( 1 )=-8.90506
X( 2 )= 18.1486
X( 3 )= 15.2132
X( 4 )=-0.752403
X( 5 )=-0.938303



DO YOU WISH TO RUN AGAIN ?YES

HOW MANY VARIABLES ARE THERE IN YOUR SYSTEM OF EQUATIONS ?3

COEFFICIENTS OF ROW 1 ?1,2,1

COEFFICIENTS OF ROW 2 ?2,1,2

COEFFICIENTS OF ROW 3 ?3,2,3

THERE IS NO UNIQUE SOLUTION TO YOUR SYSTEM OF EQUATIONS.


DO YOU WISH TO RUN AGAIN ?NO
```

3. Comments about Program (14)

a. In Step 150 the computer asks for the number of variables in your system of equations. Step 190 DIMensions matrix A by "filling out" a U x U matrix with zeros. Otherwise the computer would assume that the dimensions of A are 30 x 30 from Step 100.

Steps 200 through 270 INPUT the coefficients in each equation and convert them to entries of the matrix A. In Step 220 the coefficients of the linear system are called for, row by row, as *vectors*. Steps 230 through 250 change these U x 1 column vectors to rows of the matrix A.

Step 300 sets the matrix A equal to its own inverse. Step 310 checks for a DETerminant equal to zero. Note that a possible round-off error is taken into consideration by including ABS(DET) >= 1E-7 in the statement of Step 310.

In Step 350 B is dimensioned by filling out a U x 1 column vector with zeros. Step 370 calls for the INPUT of a column vector B. Step 390 ensures that you the user have typed the correct number of constants. If you have not, the computer tells you so (Steps 400 and 410) and then asks for a new set of constants.

Step 460 multiplies the matrix A (a U x U matrix) by B(a U x 1 matrix), and calls the resultant U x 1 matrix X. Steps 490 through 510 print the solution of the matrix equation, as you can see in the OUTPUT.

b. Notice again the use of NUM in Step 390. Recall that NUM equals the number of values entered in a MAT INPUT statement.

c. Notice in the OUTPUT that the first example shown is the solution of matrix Equation (2) given in Paragraph 1 above. The second example exhibits the solution of three equations with three variables each. You can solve these two sets of equations without too much difficulty and by nonmatrix methods. However, the third example shown, with five equations with five variables each and with rather unusual coefficients and constants, would be exceedingly difficult to solve by any method other than a computer.

Note in the last part of the OUTPUT that, when the determinant of the matrix of coefficients is equal to zero, the computer prints the message THERE IS NO UNIQUE SOLUTION TO YOUR SYSTEM OF EQUATIONS.

d. Note that the DIMension statement of Step 100 allows us to solve up to thirty equations with thirty variables each. We stipulated matrices of maximum dimensions of 30 x 30 so that we would not exceed the storage capabilities of some computer systems and because we do not anticipate solving a system of more than thirty variables.

e. Although the method of solving matrix equations as given in Program (14) utilizes the MAT statements in BASIC, it should be mentioned that there are other, more efficient and accurate methods of solving linear systems, such as the Gaussian elimination method. These other methods are beyond the scope of this book, however.

EXERCISES 8-3

1. Write a program which will find the sum of two properly dimensioned matrices without using MAT statements.

2. Write a program which will find the product of two properly dimensioned matrices without using MAT statements.

3. Write a program which will find the transpose of a matrix without using MAT statements.

4. Write a program which will graph any function such that the X-axis is horizontal and the Y-axis is vertical.

5. Write a program which will graph any parametrically defined curve.

8-4 Summary

In this chapter you first learned some of the theory of matrices, and then how to use the MATrix statements in BASIC. The theory of matrices was then combined with the use of MAT statements to write a program which can solve a linear system of n equations with n variables each.

The solution of a linear system containing four variables by a nonmatrix approach can be quite a tedious task. The solution of linear systems containing five variables by any approach is a very tedious task. Before the advent of the computer, solutions of linear systems containing six or more variables were rare. You can now solve any linear system containing up to thirty variables with ease!

CHAPTER 9

After a brief discussion of the concepts of mean, standard deviation, coefficient of correlation, and the equation for a regression line, we shall introduce you to a program in this chapter which utilizes these concepts to study the statistical relationship between two variables.

You will also learn how to write a program which finds the approximate area under the graph of any function by means of the Rectangular Rule, the Trapezoidal Rule, the Midpoint Rule, and Simpson's Rule.

We shall feel free to utilize any BASIC statement we need in order to write these programs, as you have now been exposed to all the commands available at present on most BASIC systems.

9-1 A Statistics Package: Program (15)

1. A Statistics Problem

A statistical problem arises when we have a rather large mass of data, and we want to organize that data in some intelligible way so that we can interpret it.

For example, suppose we have the results of a number of students who took several CEEB (College Entrance Examination Board) examinations. Suppose each student took (not all at the same time, of course) the SAT (Scholastic Aptitude Test − Verbal and Mathematical), along with the following achievement tests: English Composition, Mathematics Level I, Mathematics Level II, Science (Physics, Biology, or Chemistry), and a Language (Classical or Modern).

2. Statistical Relationships

We can derive equations for and calculate several values which indicate quantitative relationships in the data. Among these are the *mean, standard deviation, coefficient of correlation,* and the equation of the *regression line*. We shall discuss each of these ideas below and show how to compute the respective values for them. The presentation of these ideas must necessarily be somewhat brief in this book. If you have not studied statistics and are not familiar with these concepts, then you may wish to consult a textbook on statistics to reinforce each of these ideas.

3. The Use of Matrices to Tabulate Data

If we plan to use the computer to help us evaluate a set of data, one of the first problems which we must consider is how to handle and tabulate the data. In Chapter 8 you learned how to write a matrix and how to write a program dealing with matrices. Let us now write the data of the results of I students in J examinations (or variables) each as a matrix $A(R, C)$, where R represents the student number, from 1 to I, and C represents the examination number, from 1 to J. Thus we have an I x J array of numbers, or a matrix whose dimensions are I x J, which looks like this:

Table (5). TABULATION OF DATA

		\multicolumn{5}{c}{Examination Number}
		1	2 · · ·	C · · ·	J
Student Number	1	A(1,1)	A(1,2) · · ·	A(1,C) · · ·	A(1,J)
	2	A(2,1)	A(2,2) · · ·	A(2,C) · · ·	A(2,J)
	·	·	·	·	·
	·	·	·	·	·
	·	·	·	·	·
	R	A(R,1)	A(R,2) · · ·	A(R,C) · · ·	A(R,J)
	·	·	·	·	·
	·	·	·	·	·
	·	·	·	·	·
	I	A(I,1)	A(I,2) · · ·	A(I,C) · · ·	A(I,J)

4. The Arithmetic Mean

For each examination C we can calculate the arithmetic mean $M(C)$ by simply adding the scores of each student R (the numbers in the column) and dividing by the number of students I. We write

$$M(C) = \sum_{R=1}^{I} \frac{A(R,C)}{I}. \quad (1)$$

The symbol Σ (the Greek letter sigma) above represents the sum of A(R, C) as R is incremented by 1 from 1 to I. That is,

$$\sum_{R=1}^{I} \frac{A(R,C)}{I} = \frac{A(1,C) + A(2,C) + A(3,C) + \cdots + A(I,C)}{I}.$$

5. The Standard Deviation

The standard deviation is a number which is a measure of dispersion from the arithmetic mean of our data. That is, the standard deviation measures the amount of "spread" away from the mean.

In order to determine how much of a spread we have in the scores for a test C, we must first determine how much each value of A (R, C) deviates from the mean M (C):

$$X(R, C) = A(R, C) - M(C). \quad (2)$$

Since we want to emphasize the deviations from the mean, we square the number in equation (2) above:

$$(X(R,C))^2 = (A(R, C) - M(C))^2. \quad (3)$$

We then find the sum of the terms represented in equation (3), and divide that result by I in order to arrive at an average or mean deviation from the mean, called the variance S:

$$S = \frac{\sum_{R=1}^{I} (X(R,C))^2}{I}. \quad (4)$$

In order to obtain the same unit of measure as the one with which we started for A (R, C), we shall take the positive square root of the variance. We call that number the *standard deviation* D (C):

$$D(C) = \sqrt{\frac{\sum_{R=1}^{I} (A(R,C) - M(C))^2}{I}} \tag{5}$$

You can see that the number D (C) is the measure of the amount of deviation of the individual scores A (1, C), A (2, C), ..., A (I, C) from the mean score M (C).

6. The Coefficient of Correlation

One significant question we can ask is, "How well can the score on the English Achievement Test be predicted from the SAT Verbal score?"

In order to answer such a question, we must investigate those ideas of a coefficient of correlation between two sets of scores, and a regression line through the points which represent the data. Let us consider the correlation coefficient first.

To calculate the coefficient of correlation between two tests E and F, we first take the difference between each of the A(R, E) values and M(E):

$$X(R, E) = A(R, E) - M(E). \tag{6}$$

Similarly for A (R, F) and M (F):

$$X(R, F) = A(R, F) - M(F). \tag{7}$$

Let us find the product X(R, E) and X(R, F) for each student R, find the sum of these products, and divide by I:

$$S = \frac{\sum_{R=1}^{I} X(R,E)*X(R,F)}{I} \tag{8}$$

The number S above is the average of the products of the deviations of the A(R, E) and A(R, F) values from their respective mean values, M(E) and M(F).

Let us now divide S by the product of the standard deviations of D(E) and D(F):

$$R = \frac{S}{D(E)*D(F)} \:. \tag{9}$$

The number R is called the *coefficient of correlation* between the variables E and F.

The coefficient of correlation R is a number which varies between −1 and 1, in which R = 1 is a perfect correlation between two variables, R = 0 indicates no correlation, and −1 ⩽ R < 0 means that an inverse correlation exists between the two variables.

For example, if R = .8, a person who receives a high score on Test E most likely gets a high score on Test F, and a person who receives a low score on Test F also gets a low score on Test E. A coefficient correlation of R = −.8 indicates that a person who receives a high score on Test E most likely gets a low score on Test F, and vice versa. If R = 0, the two variables E and F are completely independent of each other.

7. The Percentage of Variance

The square of R, R^2, may be called the *percentage of variance* of variable E that can be attributed to its relationship with variable F. For example, since R = .69 between Test 1 and Test 3 in our data (See the OUTPUT of Program (15) in Paragraph 9 below), then $R^2 = (.69)^2 = .48$. We say that 48% of the variation in English Achievement scores is due to the relationship of verbal aptitude (as measured by the SAT Verbal Test) and English achievement (as measured by the CEEB English Achievement Test).

The remaining 52% must be attributed to factors other than verbal aptitude. For the given data as presented in Program (15), it appears that the SAT Verbal Aptitude Test is a rather good indicator of the CEEB English Achievement Test scores. Another study which could be conducted would be to correlate SAT scores with success in college, as measured by the grades received.

8. The Equation of the Regression Line

Let us now consider the *regression line* through the points which represent the data. Consider the graph on the following page of the data associated with the variables A(R, 1) and A(R, 3).

The straight line on Figure (20) is the line which can best be used to predict the score a student will receive on Test 3 (English Achievement) from his score on Test 1 (Verbal SAT). For example, a person who scores 700 on the Verbal SAT Test has a 48% chance of

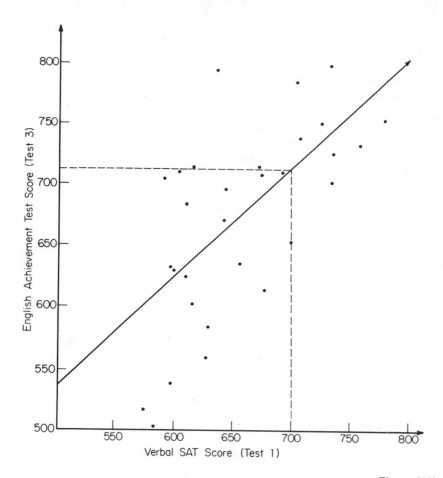

Figure (20)

scoring 715 on the English Achievement Test. The line in Figure (1) is called the regression line between the variables E (Test 1) and F (Test 3). The following discussion will show you how to find the equation of the regression line between any two variables E and F.

We know that the equation of a straight line may be written in the form:

$$Y = L * X + B, \qquad (10)$$

where L is the slope number and B is the Y-intercept number.

The slope L of the regression line for Test E predicted from Test F is dependent on the coefficient of correlation and the standard deviations of the variables E and F in the following way:

$$L = R * \frac{D(E)}{D(F)} . \qquad (11)$$

The Y-intercept B is a constant which takes into consideration the differences between the two means, as follows:

$$B = M(E) - L*M(F). \qquad (12)$$

Thus we can determine the equation of the regression line between the two variables E and F. For example, from the OUTPUT of Program (15) below, you can see that the equation of the regression line for the English Achievement Test, predicted from the SAT Verbal Test, is Y = .880534X + 97.8763.

9. Program (15): A Statistics Package

You now realize from the above discussion that calculating the mean, standard deviation, the coefficient of correlation, and the equation of the regression line is very tedious without some kind of calculating machine. However, the computer can calculate these numbers with ease, as Program (15) illustrates.

PROGRAM (15):

```
100 DIM A(50,20),X(50,20),M(20),D(20),A$(20)
110 PRINT "THIS PROGRAM COMPUTES THE MEAN AND STANDARD DEVIATION OF"
120 PRINT "A SET OF DATA, AND THE COEFFICIENT OF CORRELATION AND"
130 PRINT "THE EQUATION OF THE REGRESSION LINE FOR TWO VARIABLES."
140 REM---------READING VARIABLE (TEST) NAMES AND DATA----------
150 READ I,J
160 MAT READ A$(J),A(I,J)
170 REM---------MEANS AND STANDARD DEVIATIONS----------
180 FOR C=1 TO J
190     S=A(1,C)
200     FOR R=2 TO I
210         S=S+A(R,C)
220     NEXT R
230     M(C)=S/I
240     S=0
250     FOR R=1 TO I
260         X(R,C)=A(R,C)-M(C)
270         S=S+X(R,C)↑2
280     NEXT R
290     D(C)=SQR(S/I)
300 NEXT C
310 REM---------PRINT-OUT OF MEAN AND STANDARD DEVIATION----------
320 PRINT
330 PRINT
340 PRINT
350 PRINT " "," MEAN"," STANDARD DEVIATION"
360 PRINT
```

```
370 FOR C=1 TO J
380     PRINT A$(C),M(C),D(C)
390 NEXT C
400 REM--------CORRELATION COEFFICIENT AND REGRESSION LINE--------
410 READ E,F      'PREDICTED, PREDICTED FROM
420 IF F=-1000 THEN 950
430 S=X(1,E)*X(1,F)
440 FOR R=2 TO I
450     S=S+X(R,E)*X(R,F)
460 NEXT R
470 R=S/(I*D(E)*D(F))
480 L=R*D(E)/D(F)
490 B=M(E)-L*M(F)
500 REM----PRINT-OUT OF CORRELATION COEFFICIENT AND REGRESSION LINE----
510 PRINT
520 PRINT
530 PRINT
540 PRINT
550 PRINT "THE COEFFICIENT OF CORRELATION BETWEEN THE "A$(F)
560 PRINT "AND THE "A$(E);" IS: "R
570 PRINT
580 PRINT "THE EQUATION OF THE REGRESSION LINE FOR THE "A$(E)
590 PRINT "PREDICTED FROM THE "A$(F);" IS: Y="L;"X+"B
600 GOTO 410
610 DATA 30, 7
620 DATA ENG SAT, MATH SAT, ENG ACH., MATH I, MATH II
630 DATA SCIENCE, LANGUAGE
640 DATA 641, 730, 792, 800, 800, 732, 788
650 DATA 595, 685, 633, 725, 716, 537, 577
660 DATA 701, 748, 781, 800, 794, 745, 771
670 DATA 667, 703, 718, 800, 733, 523, 736
680 DATA 641, 703, 696, 692, 760, 644, 658
690 DATA 599, 699, 633, 792, 681, 476, 549
700 DATA 628, 748, 585, 770, 742, 535, 603
710 DATA 734, 775, 702, 800, 800, 772, 734
720 DATA 608, 703, 718, 800, 800, 658, 709
730 DATA 694, 703, 649, 781, 800, 752, 749
740 DATA 687, 748, 707, 748, 794, 772, 765
750 DATA 656, 735, 635, 737, 754, 623, 653
760 DATA 772, 645, 753, 621, 716, 576, 730
770 DATA 731, 708, 724, 661, 689, 722, 625
780 DATA 496, 645, 553, 611, 680, 539, 635
790 DATA 724, 735, 753, 760, 781, 659, 732
800 DATA 670, 654, 706, 730, 716, 578, 674
810 DATA 616, 717, 712, 690, 671, 638, 614
820 DATA 609, 726, 624, 703, 710, 616, 717
830 DATA 629, 663, 559, 690, 716, 572, 587
840 DATA 575, 672, 518, 661, 707, 567, 556
850 DATA 643, 681, 671, 700, 726, 669, 632
860 DATA 595, 728, 535, 692, 771, 771, 577
870 DATA 588, 654, 706, 700, 744, 482, 633
880 DATA 609, 672, 682, 661, 744, 589, 654
890 DATA 704, 745, 741, 730, 790, 708, 747
900 DATA 758, 800, 731, 786, 800, 778, 751
910 DATA 731, 800, 800, 721, 759, 800, 708
920 DATA 668, 800, 615, 753, 800, 778, 697
930 DATA 562, 709, 502, 800, 800, 614, 670
940 DATA 3, 1, 6, 1, 7, 1, 4, 2, 5, 2, 6, 2, 6, 5, 1, -1000
950 END
```

OUTPUT:

THIS PROGRAM COMPUTES THE MEAN AND STANDARD DEVIATION OF
A SET OF DATA, AND THE COEFFICIENT OF CORRELATION AND
THE EQUATION OF THE REGRESSION LINE FOR TWO VARIABLES.

	MEAN	STANDARD DEVIATION
ENG SAT	651.033	62.9918
MATH SAT	714.467	43.73
ENG ACH.	671.133	80.4225
MATH I	730.5	55.2037
MATH II	749.8	42.1374
SCIENCE	644.1	97.3014
LANGUAGE	674.367	68.1481

THE COEFFICIENT OF CORRELATION BETWEEN THE ENG SAT
AND THE ENG ACH. IS: .689688

THE EQUATION OF THE REGRESSION LINE FOR THE ENG ACH.
PREDICTED FROM THE ENG SAT IS: Y= .880534 X+ 97.8763

THE COEFFICIENT OF CORRELATION BETWEEN THE ENG SAT
AND THE SCIENCE IS: .558888

THE EQUATION OF THE REGRESSION LINE FOR THE SCIENCE
PREDICTED FROM THE ENG SAT IS: Y= .863296 X+ 82.0653

THE COEFFICIENT OF CORRELATION BETWEEN THE ENG SAT
AND THE LANGUAGE IS: .600813

THE EQUATION OF THE REGRESSION LINE FOR THE LANGUAGE
PREDICTED FROM THE ENG SAT IS: Y= .649994 X+ 251.199

THE COEFFICIENT OF CORRELATION BETWEEN THE MATH SAT
AND THE MATH I IS: .545554

THE EQUATION OF THE REGRESSION LINE FOR THE MATH I
PREDICTED FROM THE MATH SAT IS: Y= .688694 X+ 238.451

THE COEFFICIENT OF CORRELATION BETWEEN THE MATH SAT
AND THE MATH II IS: .59285

THE EQUATION OF THE REGRESSION LINE FOR THE MATH II
PREDICTED FROM THE MATH SAT IS: Y= .571259 X+ 341.655

THE COEFFICIENT OF CORRELATION BETWEEN THE MATH SAT
AND THE SCIENCE IS: .749775

THE EQUATION OF THE REGRESSION LINE FOR THE SCIENCE
PREDICTED FROM THE MATH SAT IS: Y= 1.66828 X+-547.834

THE COEFFICIENT OF CORRELATION BETWEEN THE MATH II
AND THE SCIENCE IS: .664212

THE EQUATION OF THE REGRESSION LINE FOR THE SCIENCE
PREDICTED FROM THE MATH II IS: Y= 1.53376 X+-505.915

---

10. Comments About Program (15)

a. The REM statements indicate what each section of the program does. Step 150 reads the number of students and the number of examinations. Step 160 first reads the name of each test C, and then assigns that name to the string variable A\$(C). Thus A\$(1) is set equal to ENG SAT, A\$(2) is set equal to MATH SAT, etc. Step 160 also reads the scores for each test, row by row, from the MAT READ statement. The loop in Steps 180 through 300 calculates the means M(C) and standard deviations D(C) for each test C, where C = 1, 2, 3, ... J, according to the discussion in Paragraphs 4 and 5 above. Steps 190 through 220 find the sum S of each score A(R, C). In accordance with Equation (1) of Paragraph 4 above, Step 230 divides each sum S by the number of students I to obtain the mean M(C).

The statement in Step 260 subtracts the mean M(C) of each variable from each test score A(R, C), and sets this difference equal to X(R, C), as in Equation (2) of Paragraph 5. Step 270 computes the sum S of the squares of X(R, C) for R = 1, 2, 3, ..., I according to Equation (4). Step 290 finds the standard deviation D(C) for each examination in accordance with Equation (5). Steps 320 through 390 print the means and standard deviations in tabular form, as you can see in the OUTPUT.

b. Steps 410 through 490 calculate the coefficient of correlation and the equation of the regression line of Test E, predicted from Test F, as discussed in Paragraphs 6 and 8 above. Steps 430 through 460 calculate the sums of the products of X(R, E) and X(R, F) according to

Equation (8) above, and Step 470 calculates the coefficient of correlation R using Equation (9). Step 480 calculates the slope L according to Equation (11), and Step 490 finds the Y-intercept B of the regression line using Equation (12). Steps 510 through 590 print the coefficient of correlation R and the equation of the regression line of Test E predicted from Test F, as you can see in the OUTPUT. Note the use of a *flag* to stop the program in Step 420.

c. Although the variables and data used in Program (15) relate to a study of CEEB scores, the program itself can be used for any statistical study which requires the calculation of the mean, the standard deviation, the coefficient of correlation, and the equation of the regression line for two variables.

Note that the DIMension statement in Step 100 allows a maximum of fifty subjects (students) and twenty variables (examinations). If this number is insufficient for your needs, then of course you can change the DIMensions to larger numbers. Bear in mind, however, that on some systems you will probably approach or exceed the capabilities of the computer.

d. The DATA of Steps 640 through 930 were taken from a random sample of actual CEEB test results of students at Roxbury Latin School over the five-year period, 1964-1969.

EXERCISES 9-1

1. Use Program (15) but your own DATA to conduct a statistical study concerning whatever interests you.

2. Write your own program for finding the mean, the standard deviation, the coefficient of correlation, and the linear regression equation for two variables X and Y.

3. Write a program which makes a graph similar to Figure (20) for any two variables, using the data in Program (15). The program should print a period for each point and should indicate the regression line with asterisks.

9-2 Area under the Curve of a Function F; Programs (16), (17)

1. Area Under the Curve of a Function F

One of the two fundamental ideas one studies in a first-year calculus course is finding the area under the curve of a function F. (The other fundamental idea is finding the slope of the tangent line to a curve at a given point on that curve. The program for this idea was given and discussed in Section 6-5).

Consider the following figure.

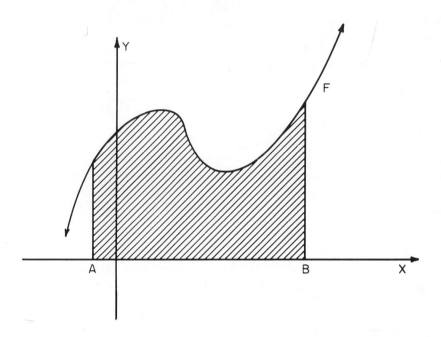

Figure (21)

Let A be any point on the X-axis, and let B be another point on the X-axis to the right of A. We are interested in calculating the area of the region enclosed by the intersection of the graphs of the inequalities $A \leqslant X \leqslant B$ and $0 \leqslant Y \leqslant F(X)$, as shown by the shaded region in Figure (21).

There are basically four methods by which we can *approximate* the area under the curve of a given function F: the Rectangular Rule, the Trapezoidal Rule, the Midpoint Rule, and Simpson's Rule. We shall discuss each of these rules in turn, and then write a program which includes all four methods.

2. The Rectangular Rule

Consider the function F whose equation is $F(X) = X^2$. Let us find an approximation of the area under the graph of F from $X = A$ to $X = B$ by the Rectangular Rule.

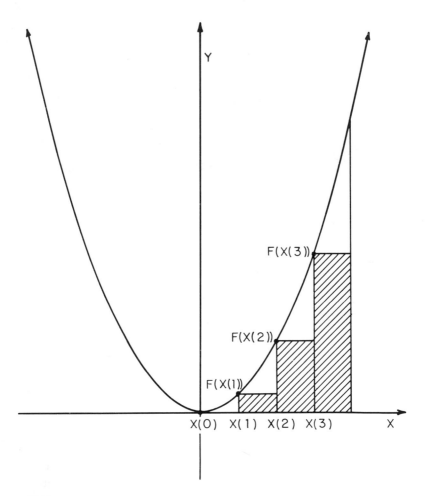

Figure (22)

Let us divide the interval from X = A to X = B into N equal parts. The length D of each of these intervals is

$$D = \frac{B - A}{N}. \tag{1}$$

Let us label the values of these intervals along the X-axis as follows:

X(0) = A, X(1) = A + D, X(2) = A + 2D, ..., X(N-1) = A+(N-1)D.

At each of the points associated with X(0), X(1), X(2),

..., X(N-1), we draw a line segment from the curve perpendicular to the X-axis, and a second line segment from these intersecting points F(X(0)), F(X(1)), F(X(2)), ..., F(X(N-1)) on the curve parallel to the X-axis to the line to the right of each point, as shown in Figure (22). We have constructed N rectangles, each of width D units, and each of length F(X(0)), F(X(1)), F(X(2)), ..., F(X(N-1)) units.

The sum A1 of the areas of these rectangles is an approximation to the area under the curve of F from X = A to X = B. If we let Y(0) = F(X(0)), Y(1) = F(X(1)), ..., Y(N-1) = F(X(N-1)), then we can find the sum A1 of the areas of the rectangles under the curve of F by the formula:

$$A1 = D*Y(0) + D*Y(1) + D*Y(2) + \cdots + D*Y(N-1)$$

$$= D*[Y(0) + Y(1) + Y(2) + \cdots + Y(N-1)]. \qquad (2)$$

For example, if A = 0, B = 2 and N = 4 for the function $F(X) = X^2$, then:

$$D = \frac{2-0}{4} = \frac{1}{2} \quad \text{and}$$

$$A1 = \frac{1}{2}\left(0 + \frac{1}{4} + \frac{4}{4} + \frac{9}{4}\right)$$

$$= \frac{7}{4}.$$

You can easily see that as the number N of subintervals increases, then the sum A1 of the areas of the rectangles approaches the actual area under the curve of F. For large values of N, the arithmetic calculations become quite tedious. However, the computer can handle these routine calculations with ease.

3. The Trapezoidal Rule

Rather than our using the Rectangular Rule and considering large values of N to find a close approximation to the actual area under the curve of F, we can begin again and construct N trapezoids. For the same value of N, the sum of the area of the trapezoids should give us a closer approximation to the actual area than the sum of the area of the rectangular regions.

9-2 Area under the Curve of a Function F; Programs (16), (17)

You know that the area T of a trapezoid is given by the formula:

$$T = \frac{D}{2}\left[Y(I) + Y(I+1)\right],$$

where Y(I) and Y(I+1) are the lengths of the bases, and D is the distance between those bases. Let us construct trapezoids on the curve of $F(X) = X^2$.

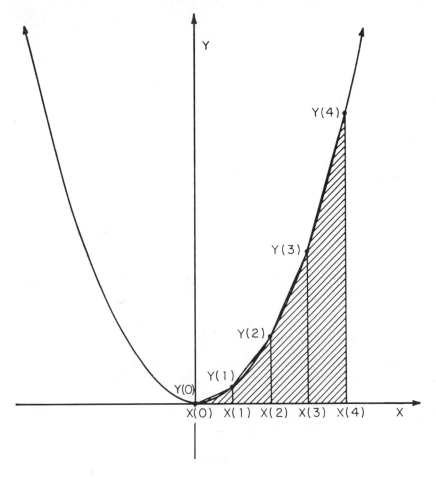

Figure (23)

The sum A2 of the areas of the trapezoids under the curve of F can be determined from the formula:

$$A2 = \frac{D}{2}*[Y(0) + Y(1)] + \frac{D}{2}*[Y(1) + Y(2)] + \cdots + \frac{D}{2}*[Y(N-1) + Y(N)]$$

$$= D*\left[\frac{1}{2}*Y(0) + Y(1) + Y(2) + \cdots + Y(N-1) + \frac{1}{2}*Y(N)\right]. \quad (3)$$

For example, if $F(X) = X^2$, $A = 0$, $B = 2$ and $N = 4$, then

$$A2 = \frac{1}{2}\left[\frac{1}{2}(0) + \frac{1}{4} + 1 + \frac{9}{4} + \frac{1}{2}(4)\right]$$

$$= \frac{1}{2}\left(\frac{11}{2}\right)$$

$$= \frac{11}{4}.$$

Again you can see that if we took a large value for N, then the trapezoidal approximation would be much closer to the actual area than the value for N = 4. Also, the trapezoidal approximation for N = 4 should be closer to the actual value for the area under the curve of $F(X) = X^2$ from $X = 0$ to $X = 2$ than the rectangular approximation for N = 4.

4. The Midpoint Rule

The Midpoint Rule gives us an even closer approximation to the actual area under a curve than both the Rectangular and Trapezoidal Rules. Consider the figure on the next page.

In Figure (24) we let Z(I) be the value of the function at the midpoint of each interval:

$$Z(I) = F((X(I) + X(I-1))/2).$$

Thus the area of each rectangle is given by the equation

$$M = D*Z(I).$$

9-2 Area under the Curve of a Function F; Programs (16), (17)

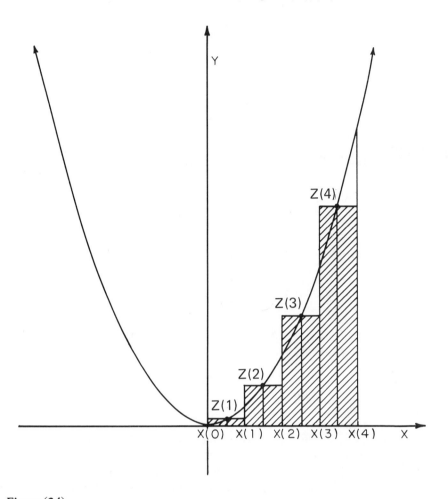

Figure (24)

The sum A3 of the areas of the rectangles constructed on the curve is given by:

$$A3 = D*Z(1) + D*Z(2) + \cdots + D*Z(N)$$
$$= D*(Z(1) + Z(2) + \cdots + Z(N)) \,. \qquad (4)$$

For example, if $F(X) = X^2$, $A = 0$, $B = 2$ and $N = 4$, then

$$A3 = \frac{1}{2}\left[\left(\frac{0+\frac{1}{2}}{2}\right)^2 + \left(\frac{\frac{1}{2}+1}{2}\right)^2 + \left(\frac{1+\frac{3}{2}}{2}\right)^2 + \left(\frac{\frac{3}{2}+2}{2}\right)^2\right]$$

$$= \frac{1}{2}\left(\frac{1}{16} + \frac{9}{16} + \frac{25}{16} + \frac{49}{16}\right)$$

$$= \frac{1}{2}\left(\frac{84}{16}\right)$$

$$= \frac{21}{8}.$$

Again you see that taking a larger value for N than N = 4 would give us a better approximation of the actual area. Also, the area from X = 0 to X = 2 by the Midpoint Rule for $F(X) = X^2$ using 4 subintervals should be closer to the actual value than the rectangular and trapezoidal approximations.

5. Simpson's Rule

A fourth method of approximating the area under the graph of a function F, which is even more accurate than the Midpoint Rule for a given value of N, was invented by Thomas Simpson (1710-1761) and bears his name. The development of Simpson's Rule is beyond the scope of this book, but the idea behind that development is quite simple.

Rather than approximating the area under a curve by a sum of rectangles or trapezoids, let us draw a parabolic arc through the points associated with Y(I), Y(I+1) and Y(I+2).

In fact, if we draw the parabolic arcs from Y(I) through Y(I+1) to Y(I+2), for I = 0 to N-2 for the graph of $F(X) = X^2$, which is itself a parabola, as in Figure (25), then we should get the *exact* area under the curve.

Simpson's Rule is as follows:

$$A4 = \frac{D}{3}*\left[Y(0)+4*Y(1)+2*Y(2)+4*Y(3)+\cdots+2*Y(N-2)+4*Y(N-1)+Y(N)\right], \quad (5)$$

for an *even* number of subintervals N.

9-2 Area under the Curve of a Function F; Programs (16), (17)

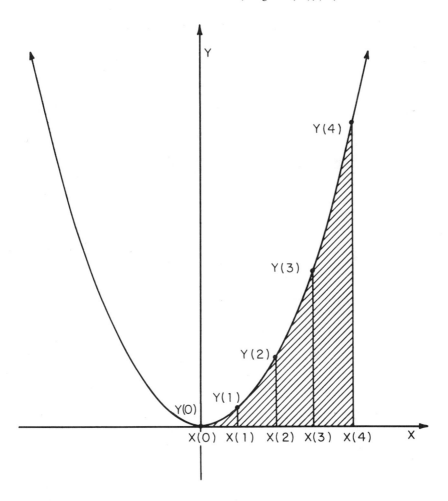

Figure (25)

For the example of $F(X) = X^2$, $A = 0$, $B = 2$ and $N = 4$, we have:

$$A4 = \frac{\frac{1}{2}}{3}\left[0 + 4\left(\frac{1}{4}\right) + 2(1) + 4\left(\frac{9}{4}\right) + 4\right]$$

$$= \frac{1}{6}(1 + 2 + 9 + 4)$$

$$= \frac{8}{3} \ .$$

If you have studied definite integration in calculus, then you know that:

$$A5 = \int_0^2 X^2 dX$$

$$= \frac{1}{3}X^3 \Big|_0^2$$

$$= \frac{1}{3}(8) - \frac{1}{3}(0)$$

$$= \frac{8}{3} .$$

Hence A4 = A5, and Simpson's Rule gives us the exact value of the area under the curve in this particular example. Actually, Simpson's Rule will give us the *exact* value of the area under the curve of any polynomial function of degree less than 4.

6. Program (16): The Area under the Curve of a Function F

We can easily write a program which approximates the area under the curve of any function FNA from X = A to X = B from the formulas developed in Paragraphs 2 through 5 above. The program is as follows.

PROGRAM (16):

```
10 DEF FNA(X)=X
20 DIM X(500),Y(500)
30 PRINT "THIS PROGRAM CALCULATES THE AREA UNDER THE CURVE OF ANY ";
40 PRINT "FUNCTION."
50 REM----------DEFINING FUNCTION----------
60 PRINT
70 PRINT "HAVE YOU DEFINED YOUR FUNCTION";
80 INPUT Q$
90 IF Q$>="Y" THEN 120
100 PRINT "TYPE '10 DEF FNA(X)=(YOUR FUNCTION)', <RETURN>, 'RUN'."
110 STOP
120 PRINT
130 PRINT
140 REM----------INTERVAL AND SUBINTERVALS----------
150 PRINT "WHAT IS THE LOWER BOUND OF YOUR INTERVAL";
```

```
160 INPUT A
170 PRINT "WHAT IS THE UPPER BOUND OF YOUR INTERVAL";
180 INPUT B
190 IF B<A THEN 150
200 PRINT "HOW MANY INTERVALS DO YOU WISH TO TAKE";
210 INPUT N
220 N=2*INT(N/2)
230 IF N<2 THEN 200
240 D=(B-A)/N
250 FOR I=0 TO N
260     X(I)=A+D*I
270     Y(I)=FNA(X(I))
280 NEXT I
290 REM----------RECTANGULAR RULE----------
300 S=Y(0)
310 FOR I=1 TO N-1
320     S=S+Y(I)
330 NEXT I
340 A1=D*S
350 REM----------TRAPEZOIDAL RULE----------
360 S=(Y(0)+Y(N))/2
370 FOR I=1 TO N-1
380     S=S+Y(I)
390 NEXT I
400 A2=D*S
410 REM----------MIDPOINT RULE----------
420 S=FNA((X(0)+X(1))/2)
430 FOR I=2 TO N
440     S=S+FNA((X(I-1)+X(I))/2)
450 NEXT I
460 A3=D*S
470 REM----------SIMPSON'S RULE----------
480 S=Y(0)+Y(N)
490 FOR I=1 TO N-1
500     IF I/2=INT(I/2) THEN 530
510     S=S+4*Y(I)
520     GOTO 540
530     S=S+2*Y(I)
540 NEXT I
550 A4=D/3*S
560 REM----------PRINT-OUT----------
570 PRINT
580 PRINT
590 PRINT
600 PRINT
610 PRINT TAB(25);"AREA UNDER CURVE OF FNA(X)"
620 PRINT TAB(25);"FROM"A;"TO"B;"USING:"
630 PRINT "NUMBER OF","RECTANGULAR","TRAPEZOIDAL","MIDPOINT","SIMPSON'S"
640 PRINT "SUBINTERVALS","RULE","RULE","RULE","RULE"
650 PRINT
660 PRINT N,A1,A2,A3,A4
670 PRINT
680 PRINT
690 PRINT
700 PRINT
710 REM----------REPEAT?----------
720 PRINT "DO YOU WISH TO RUN AGAIN";
730 INPUT R$
740 IF R$>="Y" THEN 60
750 END
```

OUTPUT:

```
THIS PROGRAM CALCULATES THE AREA UNDER THE CURVE OF ANY FUNCTION.

HAVE YOU DEFINED YOUR FUNCTION ?NO
TYPE '10 DEF FNA(X)=(YOUR FUNCTION)', <RETURN>, 'RUN'.



10 DEF FNA(X)=X↑2
RUN


THIS PROGRAM CALCULATES THE AREA UNDER THE CURVE OF ANY FUNCTION.

HAVE YOU DEFINED YOUR FUNCTION ?YES


WHAT IS THE LOWER BOUND OF YOUR INTERVAL ?0
WHAT IS THE UPPER BOUND OF YOUR INTERVAL ?2
HOW MANY INTERVALS DO YOU WISH TO TAKE ?4



                        AREA UNDER CURVE OF FNA(X)
                           FROM 0 TO 2 USING:
NUMBER OF       RECTANGULAR     TRAPEZOIDAL     MIDPOINT        SIMPSON'S
SUBINTERVALS    RULE            RULE            RULE            RULE

   4            1.75            2.75            2.625           2.66667



DO YOU WISH TO RUN AGAIN ?YES

HAVE YOU DEFINED YOUR FUNCTION ?YES


WHAT IS THE LOWER BOUND OF YOUR INTERVAL ?0
WHAT IS THE UPPER BOUND OF YOUR INTERVAL ?2
HOW MANY INTERVALS DO YOU WISH TO TAKE ?500



                        AREA UNDER CURVE OF FNA(X)
                           FROM 0 TO 2 USING:
NUMBER OF       RECTANGULAR     TRAPEZOIDAL     MIDPOINT        SIMPSON'S
SUBINTERVALS    RULE            RULE            RULE            RULE

  500           2.65867         2.66667         2.66666         2.66667



DO YOU WISH TO RUN AGAIN ?YES

HAVE YOU DEFINED YOUR FUNCTION ?NO
TYPE '10 DEF FNA(X)=(YOUR FUNCTION)', <RETURN>, 'RUN'.
```

9-2 Area under the Curve of a Function F; Programs (16), (17)

```
10 DEF FNA(X)=COS(2*X)+LOG(X)+0.5
RUN
```

THIS PROGRAM CALCULATES THE AREA UNDER THE CURVE OF ANY FUNCTION.

HAVE YOU DEFINED YOUR FUNCTION ?YES

WHAT IS THE LOWER BOUND OF YOUR INTERVAL ?2.1342
WHAT IS THE UPPER BOUND OF YOUR INTERVAL ?4.64289
HOW MANY INTERVALS DO YOU WISH TO TAKE ?20

```
                        AREA UNDER CURVE OF FNA(X)
                        FROM 2.1342 TO 4.64289 USING:
NUMBER OF     RECTANGULAR    TRAPEZOIDAL    MIDPOINT      SIMPSON'S
SUBINTERVALS  RULE           RULE           RULE          RULE

20            4.76029        4.77387        4.77847       4.77695
```

DO YOU WISH TO RUN AGAIN ?YES

HAVE YOU DEFINED YOUR FUNCTION ?YES

WHAT IS THE LOWER BOUND OF YOUR INTERVAL ?2.1342
WHAT IS THE UPPER BOUND OF YOUR INTERVAL ?4.64289
HOW MANY INTERVALS DO YOU WISH TO TAKE ?500

```
                        AREA UNDER CURVE OF FNA(X)
                        FROM 2.1342 TO 4.64289 USING:
NUMBER OF     RECTANGULAR    TRAPEZOIDAL    MIDPOINT      SIMPSON'S
SUBINTERVALS  RULE           RULE           RULE          RULE

500           4.77639        4.77693        4.77694       4.77693
```

DO YOU WISH TO RUN AGAIN ?YES

HAVE YOU DEFINED YOUR FUNCTION ?NO
TYPE '10 DEF FNA(X)=(YOUR FUNCTION)', <RETURN>, 'RUN'.

```
10 DEF FNA(X)=EXP(-X↑2/2)/SQR(2*3.14159)
RUN
```

THIS PROGRAM CALCULATES THE AREA UNDER THE CURVE OF ANY FUNCTION.

```
HAVE YOU DEFINED YOUR FUNCTION ?YES


WHAT IS THE LOWER BOUND OF YOUR INTERVAL ?-10
WHAT IS THE UPPER BOUND OF YOUR INTERVAL ?10
HOW MANY INTERVALS DO YOU WISH TO TAKE ?250



                         AREA UNDER CURVE OF FNA(X)
                            FROM-10 TO 10 USING:
NUMBER OF      RECTANGULAR   TRAPEZOIDAL   MIDPOINT      SIMPSON'S
SUBINTERVALS   RULE          RULE          RULE          RULE

  250             1.            1.            1.            1.


DO YOU WISH TO RUN AGAIN ?NO
```

7. Comments about Program (16)

a. The REM statements separate the program into parts, with each part performing a specific task. Steps 60 through 130 are used to define the function (recall that we discussed this method in Section 6-4). Steps 150 through 280 ask for values of A, B and N, and then compute and store the values of X(I) and Y(I) in Steps 260 and 270.

The areas A1, A2, A3 and A4 are calculated in Steps 300 through 550 according to Equations (2), (3), (4) and (5) respectively of Paragraphs 2 through 5 above. The areas are then printed in Steps 570 through 700, and Steps 720 through 740 contain statements which ask if you wish to repeat the program.

b. Study Equation (5) in Paragraph 5 above (Simpson's Rule). Note that the coefficient of Y(I) is 4 if I is odd and is 2 if I is even. Now look at Step 500. If I is even, the computer jumps to Step 530 and increments the sum S by 2*Y(I). If I is odd, then the computer continues to Step 510 and increments S by 4*Y(I). Step 550 contains the equivalent of Equation (5) of Paragraph 5 to find A4.

c. Note in Step 220 that the statement $N = 2*INT(N/2)$ always sets the value of N equal to an even integer. Thus we do not need to specify that the number of intervals N must be an even integer in this program (because of Simpson's Rule).

d. Notice in the first RUN of the OUTPUT that the values of the areas under the graph of $F(X) = X^2$ for $A = 0$, $B = 2$, and $N = 4$ are the same as those given in Paragraphs 2 through 5 above. Also notice in the second RUN of the OUTPUT that, when $N = 500$ for the same function, the values of the area as calculated by the Trapezoidal Rule and the Midpoint Rule are almost the same as the exact area of 2.66667, correct to 6 significant figures (the difference is due to the round-off error of the computer).

e. Notice in the last RUN of the OUTPUT that the area under the curve of the normal distribution function which was plotted in Section 7-3 is exactly 1.

8. Program (17): A Second Approach to the Area under the Curve of a Function F

Look at Program (16) again. Each part of the program follows directly from the formulas developed in Paragraphs 2 through 5 above. Notice that we relied heavily on subscript notation to calculate A1, A2, A3, and A4. The DIMension statement in Step 20 reserves enough space for 1002 variables. Storing this large number of variables reduces the efficiency of the computer. Furthermore, we are restricted to a maximum of 500 subintervals; this number may not guarantee the degree of accuracy we want (especially over a large interval from A to B).

Also, in the calculation of the area by Simpson's Rule, the computer is instructed to jump to either Step 530 or Step 540 each time through the loop. If the loop must be executed 500 times, then the time required for the computer to perform this jumping process is considerably increased. If we could eliminate this jumping technique in the program, we would increase its efficiency.

Furthermore, we notice that the computer is instructed to execute similar loops for each of the four rules. In fact, Steps 310 to 330 and Steps 370 to 390 calculate the same number! Recall that each time the computer executes a loop it increments the control variable and determines whether or not the control value is less than the specified final value. Thus, if we could combine the four loops into one, we could greatly increase the efficiency of the program.

The following program illustrates a more efficient method than Program (16) of approximating the area under the curve of any function F.

PROGRAM (17):

```
10 DEF FNA(X)=X
20 PRINT "THIS PROGRAM CALCULATES THE AREA UNDER THE CURVE OF ANY ";
30 PRINT "FUNCTION."
40 REM---------DEFINING FUNCTION----------
50 PRINT
60 PRINT "HAVE YOU DEFINED YOUR FUNCTION";
70 INPUT Q$
80 IF Q$>="Y" THEN 110
90 PRINT "TYPE '10 DEF FNA(X)=(YOUR FUNCTION)', <RETURN>, 'RUN'."
100 STOP
110 PRINT
120 PRINT
130 REM----------INTERVAL AND SUBINTERVALS----------
140 PRINT "WHAT ARE THE LOWER AND UPPER BOUNDS OF YOUR INTERVAL";
150 INPUT A,B
160 IF B<A THEN 140
170 PRINT "HOW MANY SUBINTERVALS DO YOU WISH TO TAKE";
180 INPUT N
190 N=2*INT(N/2)
200 IF N<2 THEN 170
210 REM----------CALCULATING AREA----------
220 D=(B-A)/N
230 H=D/2
240 S1=S2=S3=0
250 FOR X=A+D TO B-H STEP 2*D
260     Y=FNA(X)
270     S1=S1+FNA(X-D)+Y              'RECTANGULAR AND TRAPEZOIDAL RULE
280     S2=S2+FNA(X-H)+FNA(X+H)       'MIDPOINT RULE
290     S3=S3+4*Y+2*FNA(X+D)          'SIMPSON'S RULE
300 NEXT X
310 A1=D*S1
320 A2=D*(S1-FNA(A)/2+FNA(B)/2)
330 A3=D*S2
340 A4=D/3*(FNA(A)+S3-FNA(B))
350 REM----------PRINT-OUT----------
360 PRINT
370 PRINT
380 PRINT
390 PRINT
400 PRINT TAB(25);"AREA UNDER CURVE OF FNA(X)"
410 PRINT TAB(25);"FROM"A;"TO"B;"USING:"
420 PRINT "NUMBER OF","RECTANGULAR","TRAPEZOIDAL","MIDPOINT",
430 PRINT "SIMPSON'S"
440 PRINT "SUBINTERVALS","RULE","RULE","RULE","RULE"
450 PRINT
460 PRINT N,A1,A2,A3,A4
470 PRINT
480 PRINT
490 PRINT
500 PRINT
510 REM----------REPEAT?----------
520 PRINT "DO YOU WISH TO RUN AGAIN";
530 INPUT Q$
540 IF Q$>="Y" THEN 50
550 END
```

9-2 Area under the Curve of a Function F; Programs (16), (17)

OUTPUT:

THIS PROGRAM CALCULATES THE AREA UNDER THE CURVE OF ANY FUNCTION.

HAVE YOU DEFINED YOUR FUNCTION ?NO
TYPE '10 DEF FNA(X)=(YOUR FUNCTION)', <RETURN>, 'RUN'.

10 DEF FNA(X)=X↑2
RUN

THIS PROGRAM CALCULATES THE AREA UNDER THE CURVE OF ANY FUNCTION.

HAVE YOU DEFINED YOUR FUNCTION ?YES

WHAT ARE THE LOWER AND UPPER BOUNDS OF YOUR INTERVAL ?0,2
HOW MANY SUBINTERVALS DO YOU WISH TO TAKE ?4

```
                  AREA UNDER CURVE OF FNA(X)
                    FROM 0 TO 2 USING:
NUMBER OF    RECTANGULAR    TRAPEZOIDAL    MIDPOINT    SIMPSON'S
SUBINTERVALS RULE           RULE           RULE        RULE

4            1.75           2.75           2.625       2.66667
```

DO YOU WISH TO RUN AGAIN ?YES

HAVE YOU DEFINED YOUR FUNCTION ?YES

WHAT ARE THE LOWER AND UPPER BOUNDS OF YOUR INTERVAL ?0,2
HOW MANY SUBINTERVALS DO YOU WISH TO TAKE ?1000

```
                  AREA UNDER CURVE OF FNA(X)
                    FROM 0 TO 2 USING:
NUMBER OF    RECTANGULAR    TRAPEZOIDAL    MIDPOINT    SIMPSON'S
SUBINTERVALS RULE           RULE           RULE        RULE

1000         2.66267        2.66667        2.66666     2.66666
```

DO YOU WISH TO RUN AGAIN ?YES

```
HAVE YOU DEFINED YOUR FUNCTION ?NO
TYPE '10 DEF FNA(X)=(YOUR FUNCTION)', <RETURN>, 'RUN'.



10 DEF FNA(X)=COS(2*X)+LOG(X)+0.5
RUN



THIS PROGRAM CALCULATES THE AREA UNDER THE CURVE OF ANY FUNCTION.

HAVE YOU DEFINED YOUR FUNCTION ?YES


WHAT ARE THE LOWER AND UPPER BOUNDS OF YOUR INTERVAL ?2.1342,4.64289
HOW MANY SUBINTERVALS DO YOU WISH TO TAKE ?20
```

	AREA UNDER CURVE OF FNA(X)			
	FROM 2.1342 TO 4.64289 USING:			
NUMBER OF	RECTANGULAR	TRAPEZOIDAL	MIDPOINT	SIMPSON'S
SUBINTERVALS	RULE	RULE	RULE	RULE
20	4.76029	4.77387	4.77847	4.77695

```
DO YOU WISH TO RUN AGAIN ?YES

HAVE YOU DEFINED YOUR FUNCTION ?YES


WHAT ARE THE LOWER AND UPPER BOUNDS OF YOUR INTERVAL ?2.1342,4.64289
HOW MANY SUBINTERVALS DO YOU WISH TO TAKE ?1000
```

	AREA UNDER CURVE OF FNA(X)			
	FROM 2.1342 TO 4.64289 USING:			
NUMBER OF	RECTANGULAR	TRAPEZOIDAL	MIDPOINT	SIMPSON'S
SUBINTERVALS	RULE	RULE	RULE	RULE
1000	4.77666	4.77694	4.77694	4.77694

```
DO YOU WISH TO RUN AGAIN ?YES
```

HAVE YOU DEFINED YOUR FUNCTION ?NO
TYPE '10 DEF FNA(X)=(YOUR FUNCTION)', <RETURN>, 'RUN'.

10 DEF FNA(X)=EXP(-X↑2)
RUN

THIS PROGRAM CALCULATES THE AREA UNDER THE CURVE OF ANY FUNCTION.

HAVE YOU DEFINED YOUR FUNCTION ?YES

WHAT ARE THE LOWER AND UPPER BOUNDS OF YOUR INTERVAL ?0,1
HOW MANY SUBINTERVALS DO YOU WISH TO TAKE ?10000


```
                        AREA UNDER CURVE OF FNA(X)
                          FROM 0 TO 1 USING:
NUMBER OF     RECTANGULAR   TRAPEZOIDAL   MIDPOINT      SIMPSON'S
SUBINTERVALS  RULE          RULE          RULE          RULE

  10000       0.746855      0.746823      0.746823      0.746823
```


DO YOU WISH TO RUN AGAIN ?YES

HAVE YOU DEFINED YOUR FUNCTION ?NO
TYPE '10 DEF FNA(X)=(YOUR FUNCTION)', <RETURN>, 'RUN'.

10 DEF FNA(X)=EXP(-X)*COS(X)
RUN

THIS PROGRAM CALCULATES THE AREA UNDER THE CURVE OF ANY FUNCTION.

HAVE YOU DEFINED YOUR FUNCTION ?YES

WHAT ARE THE LOWER AND UPPER BOUNDS OF YOUR INTERVAL ?0,10
HOW MANY SUBINTERVALS DO YOU WISH TO TAKE ?1000

	AREA UNDER CURVE OF FNA(X) FROM 0 TO 10 USING:			
NUMBER OF SUBINTERVALS	RECTANGULAR RULE	TRAPEZOIDAL RULE	MIDPOINT RULE	SIMPSON'S RULE
1000	0.505015	0.500015	0.500003	0.500007

DO YOU WISH TO RUN AGAIN ?YES

HAVE YOU DEFINED YOUR FUNCTION ?NO
TYPE '10 DEF FNA(X)=(YOUR FUNCTION)', <RETURN>, 'RUN'.

10 DEF FNA(X)=LOG(ABS(1/COS(X)))
RUN

THIS PROGRAM CALCULATES THE AREA UNDER THE CURVE OF ANY FUNCTION.

HAVE YOU DEFINED YOUR FUNCTION ?YES

WHAT ARE THE LOWER AND UPPER BOUNDS OF YOUR INTERVAL ?0,1.5
HOW MANY SUBINTERVALS DO YOU WISH TO TAKE ?1000

	AREA UNDER CURVE OF FNA(X) FROM 0 TO 1.5 USING:			
NUMBER OF SUBINTERVALS	RECTANGULAR RULE	TRAPEZOIDAL RULE	MIDPOINT RULE	SIMPSON'S RULE
1000	0.828529	0.830515	0.830511	0.830513

DO YOU WISH TO RUN AGAIN ?NO

---

9. Comments about Program (17)

a. The table on the next page indicates the values of S1 as the computer executes the loops of Steps 250 to 300.

You can see from the pattern of Table (6) that S1 will equal the sum of all the ordinate (Y) values of FNA(X) as X goes from A to B−D in steps of D. That is,

$$S1 = FNA(A) + FNA(A+D) + FNA(A+2*D) + \ldots + FNA(B-D).$$

9-2 Area under the Curve of a Function F; Programs (16), (17)

Table (6)

First Loop:

$$X = A + D$$
$$S1 = 0 + FNA(A) + FNA(A+D)$$

Second Loop:

$$X = A + 3*D$$
$$S1 = FNA(A) + FNA(A+D) + FNA(A+2*D) + FNA(A+3*D)$$

Third Loop:

$$X = A + 5*D$$
$$S1 = FNA(A) + FNA(A+D) + FNA(A+2*D) + FNA(A+3*D)$$
$$+ FNA(A+4*D) + FNA(A+5*D)$$

(Note that we used the value B−H instead of B−D in Step 250 so that the computer would not "lose" the last value of X because of a possible round-off error.) Hence, the statement A1 = D*S1 in Step 310 calculates the area under the curve of FNA by the Rectangular Rule.

In Step 320 the statement

$$A2 = D*(S1 - FNA(A)/2 + FNA(B)/2)$$

is equivalent to

$$A2 = D*(FNA(A)/2 + FNA(A+D) + FNA(A+2*D) +$$
$$\ldots + FNA(B-D) + FNA(B)/2)$$

which is the formula for the Trapezoidal Rule.

You can develop the pattern for S2 and S3 in a similar fashion, and thus convince yourself that A3 = D*S2 in Step 330 does indeed yield the formula for the Midpoint Rule, and that the formula for A4 in Step 340 generates Simpson's Rule.

b. Note that we have set Y = FNA(X) in Step 260. Also notice that we use the value FNA(X) two times in the loop, in Steps 270 and 290. To avoid repetitive calculation of the value of FNA(X) inside this loop, we assign the value FNA(X) to the variable Y in Step 260, and then use that value of Y in Steps 270 and 290.

c. If you have studied calculus, you know that the value of the definite integral $\int_A^B FNA(X)dX$ is a number which is the area under the graph of FNA as X increases from A to B. Hence Programs (16) and (17) offer a powerful tool by which we can evaluate to 6 significant figures the area under the graph of any function.

EXERCISES 9-2

1. Explain why Program (17) is more efficient than Program (16).
2. Write a program which finds the average value and area under the curve of a function by Simpson's Rule.
3. Write a program which finds the volume of a solid of revolution of a function rotated about the X-axis, using the "disk" method of evaluation.
4. Write a program which finds the volume of a solid of revolution of a function rotated about the y-axis, using the "shell" method.
5. Write a program which finds the length of a curve and the area of the surface of revolution of a function rotated about the x or y-axis.
6. Using the Taylor series

$$SIN(X) = X - \frac{X^3}{3!} + \frac{X^5}{5!} - \frac{X^7}{7!} + \cdots + \frac{(-1)^{n-1} X^{(2n-1)}}{(2n-1)!} + \cdots ,$$

where n is the number of the term in the expansion, define a function FNS(X) whose value is SIN(X). Do *not* calculate $(2n-1)!$ and $X^{(2n-1)}$ for each term of the expansion. Rather, determine by what factor consecutive terms differ, i.e., the factor one must multiply the nth term by to obtain the (n+1)st term.

Be certain that your function FNS(X) is accurate for large values of X by comparing it with the SIN(X) function of the BASIC language. If your function is not accurate for large values of X, reduce X to a number whose absolute value is not greater than $\frac{\pi}{2}$ before using X in the above series.

9-3 Summary

In this chapter you have learned how to write a program which enables the computer to organize and tabulate a large mass of data as it computes the mean and standard deviation for a set of values of a variable, and the coefficient of correlation and linear regression equation between two sets of variables. The program is designed to handle a maximum of 50 subjects and 20 variables. By changing one DATA statement in the program, you can have the computer correlate any two of those variables.

You then learned how to write a program for finding the approximate area under the curve of any function using four different methods: The Rectangular Rule, the Trapezoidal Rule, the Midpoint Rule, and Simpson's Rule. You saw a second, more efficient program which calculated the area under the curve of any function using the same four methods, but which did not follow directly from the mathematical formulas.

You can now appreciate the full power of the computer as it solved these two unrelated but rather sophisticated problems. You should now appreciate also the language which enables us to take advantage rather easily of the power of the computer: the BASIC language.

CHAPTER 10

Summary of BASIC Commands

I. A. Special Symbols

The following symbols are used in this summary:

A, B, C are ...... matrices
e is an ...... expression
N is a ...... step number ($0 \leq N \leq 99999$)
R is a ...... relation ($=, <, >, <=, >=, <>$)
S is a ...... string
U is an ...... unsubscripted numerical variable
V is a ...... numerical variable subscripted or unsubscripted
V$ is a ...... string variable subscripted or unsubscripted

B. Definitions

1. A *numerical variable* is a letter, or a letter followed by a single digit. A, B, C, ..., Z, A0, A1, A2, ..., B0, B1, ..., Z8, Z9 are the 26 + 260 = 286 permitted numerical variables.
2. A *subscripted variable* is a single letter followed by one or two numbers or expressions contained in parentheses, e.g., A(2), D(7+3,5), K(A(2,8),FNB(LOG(2*Q))). Subscripted variables such as A2(5), B4(7,1), etc. are not allowed.
3. A *number* may be written in integral, decimal, or exponential (e.g., 1.38E−14) form.
4. An *expression* is a grammatically correct series of num-

bers, numerical variables (subscripted or unsubscripted), and functions connected by operators $(+, -, *, /, \uparrow)$ and containing parentheses, e.g., 7.3, A(2), $(A+B)\uparrow C+2.3$, $2.6*(SIN(1.35)/FNA(S*X))\uparrow(B-C)$.
5. A *string* is any sequence of alphanumeric characters, e.g., HYDROGEN, JAN. 1, AB39F.

II. **Statements**

   A. **Reserving Space**

      1. DIM $V_1(15), V_2(5,7)$

   B. **Assignment Statements**

      1. LET $V_1 = e$
         LET $V_1 = V_2 = V_3 = ... = e$
      2. READ V_1, V_2, V_3
         DATA 5.3, -6.374, 4.281E-16
         RESTORE
      3. INPUT V_1, V_2, V_3

   C. **Control Statements**

      1. GO TO N
      2. GOSUB N
         RETURN
      3. IF eRe $\begin{Bmatrix} \text{THEN} \\ \text{GO TO} \end{Bmatrix}$ N
      4. ON e $\begin{Bmatrix} \text{GO TO} \\ \text{THEN} \end{Bmatrix}$ $N_1, N_2, N_3, ...$
      5. STOP
      6. END

   D. **Loops**

      1. FOR U = e_1 TO e_2 STEP e_3
         NEXT U
         FOR-NEXT Loops may be nested but must not cross.

      2. IF eRe $\begin{Bmatrix} \text{THEN} \\ \text{GO TO} \end{Bmatrix}$ N

E. Output

1. PRINT $\begin{cases} V \\ e \\ V, \\ V_1, V_2, e; \\ V_1, e_1; e_2, V_2, V_3 \\ \text{``}S_1\text{''}, \text{``}S_2\text{''}; \text{``}S_3\text{''} \\ \text{``}S=\text{''} V \\ V_1; \text{``}S_1\text{''} V_2; \text{``}S_2\text{''} e \\ \text{TAB (e); ``S'' V} \\ \text{(Carriage) RETURN} \end{cases}$

F. Comments

1. REM
2. Apostrophe (')

III. Functions

A. Built-in Functions

$\left. \begin{array}{l} \text{ABS} \\ \text{SGN} \\ \text{INT} \\ \text{SQR} \\ \text{LOG} \\ \text{EXP} \\ \text{SIN} \\ \text{COS} \\ \text{TAN} \\ \text{COT} \\ \text{ATN} \\ \text{RND} \end{array} \right\}$ (e)

RND (or RND) and RANDOM (or RANDOMIZE)

B. **Matrix Statements**

$$\text{MAT} \begin{cases} \text{READ } A,B,...,C(I,J),... \\ \text{PRINT } A, B;C,... \\ C = A+B \\ C = A-B \\ B = (e)*A \\ C = A*B \\ B = \text{TRN}(A) \\ B = \text{INV}(A) \text{ and DET} \\ A = \text{ZER or } A = \text{ZER}(I,J) \\ A = \text{CON or } A = \text{CON}(I,J) \\ A = \text{IDN or } A = \text{IDN}(I,I) \\ \text{INPUT } A \text{ and NUM} \end{cases}$$

C. **Programmer-defined Functions**

1. DEF FNA($V_1, V_2, V_3,...$) = e
There are twenty-six functions possible: FNA, FNB, ..., FNZ. The definition may contain variables other than the arguments.
2. DEF FNA($V_1, V_2,...$)

   .
   .
   .
   LET FNA = e
   .
   .
   .
   FNEND

IV. **String Manipulation**

A. **Definition**

1. A *string variable* is a single letter (or a single letter followed by a digit) and a dollar sign (e.g., A$, X2$). A subscripted string variable has the form of a single letter followed by a dollar sign and a single subscript enclosed in

parentheses (e.g., C$(2), F$(5)). Subscripted string variables such as C1$(2), F6$(5) and L$(3,4) are not allowed.

B. **Statements**

1. Reserving Space
   a. DIM $V_1\$(15), V_1(3,4), V_2(12), V_2\$(26)$
2. Assignment Statements

   a. LET $V_1\$ = \begin{cases} \text{"S"} \\ V_2\$ \end{cases}$

   LET $V_1\$ = V_2\$ = V_3\$ = ... = \begin{cases} \text{"S"} \\ V\$ \end{cases}$

   b. READ $V_1\$, V_2\$, V_1, V_2, V_3\$$
   DATA 10,XYZ,"JAN.1,1984",2, "4TH"
   RESTORE $ (string data only)
   RESTORE * (numerical data only)
   RESTORE (both numerical and string data)

   c. INPUT $V_1\$, V_2\$, V_1, V_2, V_3\$, V_3$
   d. MAT INPUT V$
   e. MAT READ $V_1\$(5), V_2\$, V(I, J)$
3. Output
   a. PRINT "S" V$; e
   b. MAT PRINT $V_1\$, V_2\$; V_3\$$
4. Alphabetical Comparisons
   a. IF "S" R V$ THEN N
      IF V$ R "S" THEN N
      IF $V_1\$ R V_2\$$ THEN N
5. Single Character Manipulation

   a. CHANGE $\begin{cases} V\$ \text{ TO U} \\ U \text{ TO } V\$ \end{cases}$

V. **Operators**

The following are given in order of precedence. Operators on the same line are executed from left to right. Parentheses may be used to alter this order of execution.

1. ↑
2. * /
3. + −

VI. A. EDITing Characters

1. Shift O (←) deletes a previously typed character except (carriage) RETURN. ← ← deletes two previous characters, etc.
2. ALTMODE (ESCAPE or PREFIX) deletes an entire line.
3. N (carriage) RETURN deletes Step N.

CHAPTER 11

This chapter contains a list of system and EDITing commands, information about program size restrictions, and a list of ERROR messages. You should become familiar with the system and EDITing commands, as they will prove useful in programming. The list of ERROR messages is included for your reference.

11-1 System Commands

Many BASIC systems include several commands which are given with no step number. We shall call such statements *system commands*. Most computers type READY after executing these commands. The following is a list of the more useful system commands with a brief description of each.

Command	Description
NEW	Creates an empty working storage area. When you type a program ON LINE or feed a tape of your program into the computer, the instructions are stored in the working storage area. They are stored temporarily and are erased when you LOG OUT, type NEW, or type SCRATCH. The command NEW can be typed any time during an ON LINE session. When NEW ® is typed, the computer responds with NEW PROGRAM NAME--. You then type a maximum of six characters as the name of your new program.
OLD	Loads a previously stored file from the

11-1 Systems Commands

permanent storage area to the working storage area. Can be typed any time during an ON LINE session. When OLD ® is typed, the computer responds with OLD PROGRAM NAME--. You type the name under which you originally stored the program.

RUN — Instructs the computer to compile and execute the program in the working storage area.

LIST — Produces a PRINT-OUT of the entire current program in the working storage area. A heading which includes the name of the program, the time, and the date is typed at the beginning of the PRINT-OUT.

LIST--N — Produces a PRINT-OUT of the current program in the working storage area, beginning with Step N and continuing through to the end of the program.

LISTNH — Produces a PRINT-OUT of the entire current program in the working storage area, but with no heading.

LISTNH--N — Produces a PRINT-OUT of the current program in the working storage area, with no heading, beginning with Step N. (Some systems do not require that you type the two dashes.)

LENGTH — Prints the number of characters in the program which is in the working storage area.

SAVE — Files in the permanent storage area under the current program name all steps which exist in the working storage area.

REPLACE — Files in the permanent storage area the program currently in the working storage area and erases the program previously

	stored under the same program name. If you saved a program, modified it, and now wish to replace the stored program with the new version, type REPLACE. (Some systems require that you type SAVE instead of REPLACE.)
UNSAVE	Deletes from the permanent storage area the entire program which has the same name as the program currently in the working storage area.
RENAME (Program name)	Changes the name of the program in the working storage area to the name following the word RENAME. If no program name follows the word RENAME, the computer asks you for the NEW PROGRAM NAME.[1]
SCRATCH	Deletes from the working storage area the current program, without changing the name of the program.[1]
CATALOG	Lists the names of the programs in the permanent storage area filed under your account number.
STOP	Causes the computer to stop whatever it is doing.[2]
CONTROL-SHIFT P	Holding down the CONTROL and SHIFT keys and hitting the P key immediately stops the computer.[2]
TAPE	Instructs the computer to accept statements from the tape reader mechanism. (Most systems will accept a tape without this command.)

---

[1] Typing the commands RENAME and SCRATCH consecutively in either order is equivalent to typing NEW.
[2] STOP and CONTROL–SHIFT P are not used on all systems. If these keys do not stop the computer, depress that key which controls your system.

11-2 EDIT Commands

EDIT commands are also available on most systems. These system commands enable you to manipulate a program in a variety of ways. The EDIT commands are given with no step number, and the computer types READY after executing such a command. (On some systems, the word EDIT is not required before the command.)

A list of the more useful EDIT commands, with a brief description of each, is as follows.

Command	Description
EDIT DELETE N1, N2-N3, N4, ...	Deletes specific steps from the current program in the working storage area. EDIT DELETE N1, N2-N3, N4 deletes Step N1, Steps N2 through N3, and Step N4 respectively. One command may contain as many step numbers as will fit on one line.
EDIT EXTRACT N1, N2-N3, N4, ...	Deletes all steps of the current program in the working storage area *except* those specified in the parameters.
EDIT LIST N1, N2-N3, N4, ...	Lists specific steps of the current program in the working storage area. EDIT LIST N1 lists Step N1. EDIT LIST N2-N3 lists all steps from N2 to N3. EDIT LIST N3-N2 lists all steps from N3 to N2 in reverse order, if $N3 > N2$. If no parameters are given in the command, then the program is listed in reverse order.
EDIT MERGE A1, A2,N1,A3,N2, ...	Permits you to combine SAVEd programs without resequencing the step numbers of the MERGEd program. A1 is the name of the main program. The program named A2 is inserted after Step N1 of the main program A1, the program named A3 is inserted after Step N2 of the main program, etc. The step numbers of the MERGEd program are automatically resequenced beginning with Step 100 and continuing in increments of 10. For

a permanent file, the MERGEd program must be SAVEd.

EDIT MOVE N1-N2,N3   Inserts the steps between N1 and N2 inclusive after Step N3 and automatically deletes Steps N1 through N2 from their original position. The program is automatically resequenced if necessary.

EDIT PAGE A1,N,A2, A3,...   Lists programs in a paged format. A1, A2,... are the names of SAVEd programs. N is the starting page number. (If no N is included, the computer assumes that the starting page number is 1.) The programs A1, A2,... are listed in paged format, 50 lines per page. Sufficient space is left between pages for cutting the paper. At the top of each page, the appropriate page number is printed.

EDIT RESEQUENCE N1,N2,N3   Resequences the step numbers of the current program in the working storage area. The number of Step N2 is changed to Step N1, and all subsequent steps are changed in increments of N3. All references to step numbers (in GOTO, IF-THEN or ON-GOTO statements) are also automatically changed to the appropriate new step number. The value of 10 is given to N3 if it is not included in the command. Resequencing starts at the beginning of the program if N2 is not given. If no parameters are given, then the computer will change the first step to Step 100 and renumber all succeeding steps in increments of 10.

EDIT RUNOFF   Lists the entire text of the program in the working storage area without step numbers. Each line contains as many statements of the program as will fit on that line. The computer automatically inserts proper spacing between words on each line in order to create both a left and right hand margin. The

11-2 EDIT Commands 269

computer will type any text in this neat format provided each line of the original text is preceded by a step number.

EDIT WEAVE A1, A2,A3,...   Combines two or more SAVEd programs into one program under the current program name. A1, A2, A3,... are the names of the SAVEd programs. The separate programs must not contain the same step numbers. You must type SAVE if you wish to permanently store the WEAVEd program.

11-3 Size Restrictions

BASIC has no specific restrictions, in terms of numbers of steps, for a program. However, certain combinations of statements, variables, and constants may produce a program which exceeds the computer's memory capacity. The following list indicates these restrictions.

Element	Restriction
Program Length	Generally, no longer than 8 feet of typed program on Teletype paper.
Constants	No definite restriction.
Data	No definite restriction.
Lists and Tables (Vectors and Matrices)	No definite restriction.
GO TO and IF-THEN Statements	No definite restriction.
FOR-NEXT Statement	Number of nestings cannot exceed 20.
DEF and GOSUB Statements	Any combination of DEF and GOSUB statements, including nestings, cannot exceed 100.

11-4 ERROR Messages

The BASIC language has a number of ERROR messages which inform you of mistakes in your program. Some of these ERROR messages are given during compilation (after the computer has been instructed to RUN, but before actual execution of the program), and others are given during execution. Many ERROR messages include the step number in which the error occurs.

The following alphabetical list of ERROR messages, with a brief description of each and suggestions for correcting the mistakes, is given in order to help you debug your programs.

Compilation Errors

ERROR Message	Description and Correction
CUT PROGRAM OR DIMS	The program is too long, or the space reserved by the DIM statement is too large, or a combination of both. Shorten the program by reducing labels, lists, or variables. If your program contains subscripted variables with less than 11 values, include DIM statements such as DIM A(5) or DIM B(4, 3) to leave more space for your program.
DIMENSION TOO LARGE AT N	The number of elements or components of an array is too large. Reduce the size of the array or the number of DIMensioned variables.
END IS NOT LAST	Either the END statement does not have the highest step number, or there are two or more END statements in the program.
EXPRESSION TOO COMPLICATED IN N	There are too many operations or parentheses in Step N. Write two or more simpler statements.
FOR'S NESTED TOO DEEPLY AT N	There are too many FOR statements before the corresponding NEXT statements. Rearrange steps to avoid loop crossing.

11-2 EDIT Commands 271

FOR WITHOUT NEXT IN N	A NEXT statement is missing. Check that the control variable in the FOR statement is the same as the variable in the NEXT statement.
ILLEGAL CHARACTER IN N	Check for illegal character in Step N and retype Step N correctly.
ILLEGAL CONSTANT IN N	A number has more than 9 digits, or is in incorrect form, or is out of the range of the system. Check for commas in numbers (A = 1,000) or typing errors (B = 1R12).
ILLEGAL FORMAT IN N	An instruction is in unacceptable form. Check Step N for typing error.
ILLEGAL FORMULA IN N	Illegal operations, variable names, or missing parentheses. Check matching of parentheses and inclusion of multiplication symbol (*).
ILLEGAL INSTRUCTION IN N	Step N contains an instruction not included in BASIC. Check for typographical errors.
ILLEGAL LINE NUMBER AFTER N	The line number following Step N is not in correct form. Check that the line number does not contain a letter, is less than 6 digits, and is a nonnegative integer.
ILLEGAL LINE REFERENCE IN N	A character other than a number appears in a GOTO, IF-THEN, ON-GOTO, or GOSUB statement.
ILLEGAL MAT FUNCTION IN N	There is an unacceptable matrix function in Step N. Check scalar product (MAT B= (e)*A).
ILLEGAL MAT MULTIPLE IN N	Two matrices have not been properly multiplied in Step N. MAT C = C*D is illegal.
ILLEGAL MAT TRANSPOSE IN N	A matrix has been transposed incorrectly in Step N. MAT F = TRN(F) is illegal.
ILLEGAL RELATION IN N	A relation symbol other than the six defined ones of $=, >, <, >=, <=, <>$ has been used in an IF-THEN statement.

ILLEGAL VARIABLE IN N	Step N contains an illegal variable name. Variables such as A1(C) and BD are illegal.
INCORRECT NUMBER OF ARGUMENTS IN N	The number of arguments of a referenced function must equal the number of arguments in that function's DEFinition.
INCORRECT NUMBER OF SUBSCRIPTS IN N	A matrix has been illegally defined with a single subscript, or a vector with a double subscript.
MISMATCHED STRING OPERATION IN N	Illegal attempt to combine algebraically two or more strings, or to compare a string and a number, or to assign a number to a string variable or a string to a numerical variable.
NESTED DEF	Multiline function definitions must not be nested.
NEXT WITHOUT FOR IN N	A NEXT statement appears without a FOR statement. Check that the variable name in the NEXT statement is the same as the control variable in the FOR statement.
NO END INSTRUCTION	There is no END statement in the program.
NO NUMERIC DATA	The program contains at least one READ statement calling for numeric data, but no DATA statements containing that data.
NO STRING DATA	The program contains at least one READ statement calling for string data, but no DATA statement containing strings.
TOO MANY CONSTANTS AT N	Too many constants (which are difficult for the computer to handle in binary code) are used in Step N. Use a DATA statement for the constants.
UNDEFINED LINE NUMBER N1 IN N2	The step number N1 referred to in a GOTO, IF-THEN, GOSUB, or ON-GOTO statement in Step N2 does not exist in the program.

UNDEFINED FUNCTION FNA IN N	A function is used in Step N which is not DEFined in the program.
UNFINISHED DEF	FNEND has not been included at the end of a multiline function DEFinition.

The following ERROR messages occur during the execution of the program. Some errors encountered during a RUN cause an ERROR message to be printed but do not stop the execution of the program. Other ERROR messages inform you of an error and also stop the execution of the program.

Execution Errors

ERROR Message	Description and Correction
ABSOLUTE VALUE RAISED TO POWER IN N	The operation $X \uparrow Y$ has been attempted where X is a negative number and Y is a decimal. The computer calculates $(ABS(X)) \uparrow Y$ and continues execution of the program.
DIMENSION ERROR IN N	A DIMension error exists in the MAT statement in Step N. Recall that in the product of matrices, $A_{rxc} * B_{cxp} = C_{rxp}$. Execution stops.
DIVISION BY ZERO IN N	The statement in Step N instructs the computer to divide by zero. The computer substitutes its value for positive infinity (1.70141E38 on some systems) and continues execution of the program.
EXP TOO LARGE IN N	The value of EXP(X) is outside the range of the computer. The computer substitutes its value for positive infinity and execution continues.
GOSUB NESTED TOO DEEPLY AT N	Too many GOSUBs before a RETURN. This ERROR message also occurs when subroutines are exited by GO TO, IF-THEN, or

	ON-GO TO statements instead of RETURN. Execution stops.
INPUT DATA NOT IN CORRECT FORMAT–RETYPE IT	The data entered in an INPUT statement is not in the correct format (2∗3, 3↑2, etc., are illegal in INPUT statements). After the ERROR message, the computer types another question mark and waits for you to retype your INPUT data.
LOG OF NEGATIVE NUMBER IN N	Step N contains a statement which attempts to calculate LOG(X), where X is a negative number. The computer calculates LOG(ABS(X)) and continues to execute the program.
LOG OF ZERO IN N	Step N attempts to calculate LOG(0). The computer substitutes its value for negative infinity (−1.70141E38 on some systems) and continues to execute the program.
NOT ENOUGH INPUT–ADD MORE	Not enough INPUT data. Computer waits for more DATA and then continues to execute the program.
ON EVALUATED OUT OF RANGE IN N	The integral part of the value of the variable in the ON-GOTO statement in Step N is less than 1 or greater than the number of step numbers in the statement. Execution stops.
OUT OF DATA IN N	Step N attempts to read past the end of the DATA. Execution stops.
OUT OF ROOM	The space reserved in a DIM statement is not large enough. Check for endless loops in the program. Execution stops.
OVERFLOW IN N	A number X whose absolute value is greater than the computer's largest value has been generated in Step N. The computer substitutes its value of positive or negative infinity for X and continues execution of the program.

11-4 ERROR Messages

RETURN BEFORE GOSUB IN N	A RETURN is encountered before a GOSUB. Recall that a RETURN may have a smaller step number than a GOSUB, but may not be executed before the GOSUB. Check whether a subroutine was entered with a GOTO rather than a GOSUB. Execution stops.
SQUARE ROOT OF NEGATIVE NUMBER IN N	Step N attempts to calculate SQR(X) where X is less than zero. The computer calculates SQR(ABS(X)) and continues to execute the program.
SUBSCRIPT ERROR IN N	Step N uses a subscript outside the range of the DIM statement. If no DIM statement exists in the program, then the subscript used is greater than 10. Insert a PRINT statement immediately before Step N to determine the subscript in question. Execution stops.
TOO MUCH INPUT-- EXCESS IGNORED	You entered more data than requested. The computer ignores the excess data and continues to execute the program.
UNDERFLOW IN N	A number X whose absolute value is less than the computer's smallest value has been generated in Step N. The computer substiutes zero for X and continues execution of the program.
USELESS LOOP IN N	The computer has detected a loop which does nothing in the program. Execution stops.
ZERO TO A NEGATIVE POWER IN N	Step N attempts to compute 0↑B, where B is a negative number. The computer substitutes its value of positive infinity for 0↑B (1.70141E38 on some systems) and continues to execute the program.

APPENDICES

Appendix A: Algebra Programs
  Program 1:   Average of N Numbers   279
                2:   Solution of AX+B=CX+D   279
                3.   Pairs of Factors of an Integer   280
                4.   Prime Factors of Any Integer   281
                5.   Least Common Multiple and Greatest Common Factor   282
                6.   Solution of Two Simultaneous Linear Equations   284
                7.   Completion Time of Work Problem   285
                8.   To Convert a Number from Base 10 to Any Base   286
                9.   Logarithm of N in Base B   288
              10.   Approximation of the Square Root of N   289
              11.   To Find the Rth Root of N   290
              12.   To Divide One Polynomial by Another   290
              13.   To Graph a Function with X-Axis Horizontal   292
              14.   To Graph Up to 26 Functions   293
              15.   Real Zeros of Any Function by Newton's Method   297

Appendix B: Geometry Programs
              16.   Pythagorean Triplets   300
              17.   Area of Any Regular Polygon   301
              18.   To Find the Interior Angle, Diagonals of a Regular Polygon   302

Appendix C: Trigonometry Programs
              19.   Area of Triangle, Given SAS or SSS   304

Appendix D: Analytic Geometry Programs
              20.   Equation of Parabola Through Three Points   306
              21.   Analysis of $AX^2+BXY+CY^2+DX+EY+F=0$ 308

Appendix E: Calculus Programs
- 22. To Find the Limit of Any Function ... 314
- 23. Area Under Curve and Average Value of Any Function ... 319
- 24. Volume of Any Solid of Revolution ... 322
- 25. To Calculate Sin(X) by a Power Series ... 326

Appendix F: Probability Programs
- 26. Binomial Experiment ... 329

Appendix G: Special Programs
- 27. Arranging of Words in Alphabetical Order ... 331
- 28. Electronic Configuration of Any Element ... 332
- 29. Game of Chinese War ... 334
- 30. A Meaningless Technical Report ... 336

APPENDIX A
Algebra Programs

Program 1: Average of N Numbers

```
10 DIM V(1000)
20 PRINT "WHAT NUMBERS DO YOU WANT AVERAGED ";
30 MAT INPUT V
40 IF NUM=0 THEN 120
50 S=0
60 FOR I=1 TO NUM
70     S=S+V(I)        '-----------SUM OF NUMBERS
80 NEXT I
90 PRINT "THE AVERAGE IS"S/NUM
100 PRINT
110 GOTO 20
120 END
```

OUTPUT:

```
WHAT NUMBERS DO YOU WANT AVERAGED ?1,2,3
THE AVERAGE IS 2

WHAT NUMBERS DO YOU WANT AVERAGED ?-1,0,1
THE AVERAGE IS 0

WHAT NUMBERS DO YOU WANT AVERAGED ?1.2949 3, 3.4251E2, 0.0923, 87.9E-2
THE AVERAGE IS 86.1941

WHAT NUMBERS DO YOU WANT AVERAGED ?
```

Program 2: Solution of AX+B=CX+D

```
10 PRINT "THIS PROGRAM SOLVES THE EQUATION AX+B=CX+D."
20 PRINT
30 PRINT
40 PRINT "WHAT ARE A,B,C,D ";
50 INPUT A,B,C,D
60 IF A<>C THEN 120
70 IF B=D THEN 100
```

```
80 PRINT "YOUR EQUATION HAS NO SOLUTION SET."
90 GOTO 130
100 PRINT "YOUR EQUATION HAS A UNIVERSAL SOLUTION SET."
110 GOTO 130
120 PRINT "X=";(D-B)/(A-C)
130 PRINT
140 PRINT
150 PRINT "DO YOU WANT ME TO SOLVE ANOTHER EQUATION ";
160 INPUT Q0
170 IF Q0>="Y" THEN 20
180 END
```

OUTPUT:

```
THIS PROGRAM SOLVES THE EQUATION AX+B=CX+D.


WHAT ARE A,B,C,D ?2,3,4,1
X= 1


DO YOU WANT ME TO SOLVE ANOTHER EQUATION ?YES


WHAT ARE A,B,C,D ?1,3,1,2
YOUR EQUATION HAS NO SOLUTION SET.


DO YOU WANT ME TO SOLVE ANOTHER EQUATION ?YES


WHAT ARE A,B,C,D ?1,2,1,2
YOUR EQUATION HAS A UNIVERSAL SOLUTION SET.


DO YOU WANT ME TO SOLVE ANOTHER EQUATION ?YES


WHAT ARE A,B,C,D ?2.5467,-9.75987,3.14159,-0.9879
X=-14.7455


DO YOU WANT ME TO SOLVE ANOTHER EQUATION ?NO
```

Program 3: Pairs of Factors of an Integer

```
10 PRINT "THIS PROGRAM FINDS ALL PAIRS OF FACTORS OF AN INTEGER."
20 PRINT
30 PRINT
40 PRINT "WHAT IS THE INTEGER";
50 INPUT X
60 PRINT
70 PRINT "THE PAIRS OF FACTORS ARE:"
```

Appendix A: Algebra Programs 281

```
 80 FOR A=1 TO SQR(ABS(X))
 90    IF INT(X/A)<>X/A THEN 110
100    PRINT A,X/A
110 NEXT A
120 PRINT
130 PRINT
140 PRINT "DO YOU WISH TO RUN AGAIN";
150 INPUT Q$
160 IF Q$="Y" THEN 20
170 END
```

OUTPUT:

```
THIS PROGRAM FINDS ALL PAIRS OF FACTORS OF AN INTEGER.

WHAT IS THE INTEGER?24

THE PAIRS OF FACTORS ARE:
 1             24
 2             12
 3              8
 4              6


DO YOU WISH TO RUN AGAIN?YES

WHAT IS THE INTEGER?60

THE PAIRS OF FACTORS ARE:
 1             60
 2             30
 3             20
 4             15
 5             12
 6             10


DO YOU WISH TO RUN AGAIN?NO
```

Program 4: Prime Factors of Any Integer

```
100 PRINT "THIS PROGRAM FINDS THE PRIME FACTORS OF ANY INTEGER."
110 REM---------ENTERING NUMBER----------
120 PRINT
130 PRINT
140 PRINT
150 PRINT "WHAT NUMBER DO YOU WANT FACTORED ";
160 INPUT A
170 IF ABS(A)<=1 THEN 340
180 N=INT(ABS(A))
190 REM---------FINDING AND PRINTING PRIMES----------
200 B=0
```

```
210 FOR I=2 TO N/2
220     IF N/I>INT(N/I) THEN 300
230     B=B+1
240     IF B>1 THEN 260
250     PRINT "THE PRIME FACTORS OF"N;"ARE:"
260     PRINT I;
270     N=N/I
280     IF N=1 THEN 120
290     I=I-1
300 NEXT I
310 IF N<>INT(A) THEN 120
320 PRINT "********"N;"IS PRIME.********"
330 GOTO 130
340 END
```

OUTPUT:

```
THIS PROGRAM FINDS THE PRIME FACTORS OF ANY INTEGER.


WHAT NUMBER DO YOU WANT FACTORED ?6
THE PRIME FACTORS OF 6 ARE:
 2  3


WHAT NUMBER DO YOU WANT FACTORED ?16
THE PRIME FACTORS OF 16   ARE:
 2  2  2  2


WHAT NUMBER DO YOU WANT FACTORED ?91
THE PRIME FACTORS OF 91   ARE:
 7  13


WHAT NUMBER DO YOU WANT FACTORED ?23
******** 23   IS PRIME.********


WHAT NUMBER DO YOU WANT FACTORED ?0
```

Program 5: Least Common Multiple and Greatest Common Factor

```
100 DIM A(100),B(100)
110 PRINT "THIS PROGRAM FINDS THE LEAST COMMON MULTIPLE AND GREATEST"
120 PRINT "COMMON FACTOR OF A SERIES OF INTEGERS."
130 PRINT
140 PRINT
150 PRINT "WHAT ARE THE NUMBERS";
160 MAT INPUT A
170 IF NUM=0 THEN 530
180 K=A(1)
```

```
190 REM----------FINDING MAXIMUM----------
200 FOR I=1 TO NUM
210    IF K>=A(I) THEN 230
220    K=A(I)
230 NEXT I
240 L=1
250 G=1
260 REM----------COMPUTING LCM AND GCF----------
270 FOR J=2 TO K
280    FOR I=1 TO NUM
290       B(I)=1
300    NEXT I
310    REM---DETERMINES HOW MANY J'S ARE IN EACH OF YOUR NUMBERS---
320    FOR I=1 TO NUM
330       IF A(I)/J<>INT(A(I)/J) THEN 370
340       A(I)=A(I)/J
350       B(I)=B(I)*J
360       I=I-1
370    NEXT I
380    REM-MAX NO. OF J'S IN NOS.& MIN NO. OF J'S COMMON TO ALL NOS.
390    F=B(1)
400    M=B(1)
410    FOR I=2 TO NUM
420       IF F>=B(I) THEN 440
430       F=B(I)
440       IF M<=B(I) THEN 460
450       M=B(I)
460    NEXT I
470    L=L*F       '------COMPUTES LCM-------
480    G=G*M       '------COMPUTES GCF-------
490 NEXT J
500 PRINT "THE LEAST COMMON MULTIPLE IS "L
510 PRINT "THE GREATEST COMMON FACTOR IS"G
520 GOTO 130
530 END
```

OUTPUT:

```
THIS PROGRAM FINDS THE LEAST COMMON MULTIPLE AND GREATEST
COMMON FACTOR OF A SERIES OF INTEGERS.


WHAT ARE THE NUMBERS ?3,-9,-12,15
THE LEAST COMMON MULTIPLE IS  180
THE GREATEST COMMON FACTOR IS 3


WHAT ARE THE NUMBERS ?120,90,-15,105
THE LEAST COMMON MULTIPLE IS  2520
THE GREATEST COMMON FACTOR IS 15


WHAT ARE THE NUMBERS ?45,-24,72
THE LEAST COMMON MULTIPLE IS  360
THE GREATEST COMMON FACTOR IS 3


WHAT ARE THE NUMBERS ?36,-212,128,64,1012
THE LEAST COMMON MULTIPLE IS  15447168
THE GREATEST COMMON FACTOR IS 4
```

```
WHAT ARE THE NUMBERS ?3,-8
THE LEAST COMMON MULTIPLE IS   24
THE GREATEST COMMON FACTOR IS  1


WHAT ARE THE NUMBERS ?1225,1715,385
THE LEAST COMMON MULTIPLE IS   94325
THE GREATEST COMMON FACTOR IS  35


WHAT ARE THE NUMBERS ?
```

Program 6: Solution of Two Simultaneous Linear Equations

```
100 PRINT "THIS PROGRAM SOLVES A SET OF 2 SIMULTANEOUS LINEAR"
110 PRINT "EQUATIONS IN X AND Y IN THE FORM:"
120 PRINT " ","AX+BY=C"
130 PRINT " ","DX+EY=F"
140 PRINT
150 PRINT
160 PRINT "WHAT ARE YOUR VALUES OF A,B,C,D,E,F ";
170 INPUT A,B,C,D,E,F
180 PRINT
190 IF B/E=C/F THEN 230
200 IF A*E-B*D<>0 THEN 260
210 PRINT "YOUR EQUATIONS ARE INCONSISTENT: NO X AND Y WILL WORK."
220 GOTO 290
230 IF A/D<>C/F THEN 260
240 PRINT "YOUR EQUATIONS ARE DEPENDENT: ANY X AND Y WILL WORK."
250 GOTO 290
260 PRINT "THE SOLUTION IS:"
270 PRINT "X=";(C*E-B*F)/(A*E-B*D)
280 PRINT "Y=";(A*F-C*D)/(A*E-B*D)
290 PRINT
300 PRINT
310 PRINT "DO YOU WANT ME TO SOLVE ANOTHER SET OF EQUATIONS ";
320 INPUT Q$
330 IF Q$>="Y" THEN 140
340 END
```

OUTPUT:

```
THIS PROGRAM SOLVES A SET OF 2 SIMULTANEOUS LINEAR
EQUATIONS IN X AND Y IN THE FORM:
            AX+BY=C
            DX+EY=F


WHAT ARE YOUR VALUES OF A,B,C,D,E,F  ?1,1,1,-1,1,1

THE SOLUTION IS:
X= 0
Y= 1


DO YOU WANT ME TO SOLVE ANOTHER SET OF EQUATIONS  ?YES
```

```
WHAT ARE YOUR VALUES OF A,B,C,D,E,F   ?1,1,1,1,1,2
YOUR EQUATIONS ARE INCONSISTENT: NO X AND Y WILL WORK.

DO YOU WANT ME TO SOLVE ANOTHER SET OF EQUATIONS  ?YES

WHAT ARE YOUR VALUES OF A,B,C,D,E,F   ?5,5,5,5,5,5
YOUR EQUATIONS ARE DEPENDENT: ANY X AND Y WILL WORK.

DO YOU WANT ME TO SOLVE ANOTHER SET OF EQUATIONS  ?YES

WHAT ARE YOUR VALUES OF A,B,C,D,E,F   ?23.45,57.23,59.45,2.34,1.6,49
THE SOLUTION IS:
X= 23.0951
Y=-10.464


DO YOU WANT ME TO SOLVE ANOTHER SET OF EQUATIONS  ?NO
```

Program 7: Completion Time of Work Problem

```
10 DIM A(100)
20 PRINT "THIS PROGRAM COMPUTES THE COMPLETION TIME FOR JOBS WITH"
30 PRINT "SEVERAL MEN WORKING AT THE SAME TIME (IT ASSUMES THEY DO"
40 PRINT "NOT INTERFERE WITH EACH OTHER)."
50 PRINT "IF YOU WISH TO USE NEGATIVE TIME YOU MAY.  FOR EXAMPLE"
60 PRINT "A DRAIN CAN FILL A POOL IN SO MANY NEGATIVE HOURS."
70 PRINT
80 PRINT
90 PRINT
100 PRINT "HOW LONG WOULD THE JOB TAKE EACH MAN ALONE";
110 MAT INPUT A
120 Y=0
130 FOR I=1 TO NUM
140     IF A(I)<>0 THEN 160
150     A(I)=1E-6
160     Y=Y+1/A(I)
170 NEXT I
180 PRINT "THE COMPLETION TIME IS"1/Y
190 PRINT
200 PRINT
210 PRINT "DO YOU WISH TO RUN AGAIN";
220 INPUT Q$
230 IF Q$="Y" THEN 70
240 END
```

OUTPUT:

```
THIS PROGRAM COMPUTES THE COMPLETION TIME FOR JOBS WITH
SEVERAL MEN WORKING AT THE SAME TIME (IT ASSUMES THEY DO
NOT INTERFERE WITH EACH OTHER).
IF YOU WISH TO USE NEGATIVE TIME YOU MAY.  FOR EXAMPLE
A DRAIN CAN FILL A POOL IN SO MANY NEGATIVE HOURS.


HOW LONG WOULD THE JOB TAKE EACH MAN ALONE?1,2
THE COMPLETION TIME IS 0.666667


DO YOU WISH TO RUN AGAIN?YES


HOW LONG WOULD THE JOB TAKE EACH MAN ALONE?2,3
THE COMPLETION TIME IS 1.2


DO YOU WISH TO RUN AGAIN?YES


HOW LONG WOULD THE JOB TAKE EACH MAN ALONE?-1,-5,2,-2.5,3
THE COMPLETION TIME IS-1.30435


DO YOU WISH TO RUN AGAIN?NO
```

Program 8: To Convert a Number from Base 10 to Any Base

```
100 DIM C(100)
110 REM------------DESCRIPTION-------------
120 PRINT "THIS PROGRAM CONVERTS A NUMBER FROM BASE 10 TO ANY BASE."
130 PRINT "IF YOU ARE CONVERTING YOUR NUMBER TO A BASE GREATER THAN"
140 PRINT "10, I WILL USE THE FOLLOWING CODE IN PRINTING DIGITS"
150 PRINT "GREATER THAN 9:"
160 PRINT
170 A(0)=1
180 FOR I=58 TO 95
190     A(1)=I
200     CHANGE A TO A$
210     PRINT A$;" FOR";I-48,
220 NEXT I
230 PRINT
240 PRINT
250 PRINT
260 PRINT "WHAT NUMBER DO YOU WANT CONVERTED";
270 INPUT N
280 PRINT "IN WHAT BASE WOULD YOU LIKE THE NUMBER";
290 INPUT B
300 B=INT(B)
```

```
310 IF B<2 THEN 280
320 IF B<48 THEN 350
330 PRINT "I CAN'T HANDLE BASES THAT BIG."
340 GOTO 640
350 PRINT
360 REM----------CONVERTING NUMBER----------
370 PRINT N;"IN BASE"B;"IS ";
380 IF N>=0 THEN 400
390 PRINT "-";
400 N=ABS(N)
410 E=1
420 IF N<1 THEN 460
430 REM--FINDING HIGHEST POWER OF BASE LESS THAN NUMBER--
440 FOR E=-10 TO 1000
450     IF B↑E>N THEN 480
460 NEXT E
470 REM----------CONVERTING NUMBER----------
480 C(0)=P=1
490 FOR H=E-1 TO -10 STEP -1
500     C(C(0))=INT(N/B↑H)+48       'CODE NUMBER FOR DIGITS
510     N=N-INT(N/B↑H)*B↑H
520     IF P=1 THEN 540
530     IF ABS(N)<1E-7 THEN 610
540     C(0)=C(0)+1
550     IF H<>1 THEN 570
560     P=0     'FLAG FOR DECIMAL POINT
570     IF H<>0 THEN 600
580     C(C(0))=46      'CODE NUMBER FOR DECIMAL POINT
590     C(0)=C(0)+1
600 NEXT H
610 CHANGE C TO N$
620 PRINT N$
630 REM------------REPEAT ?-------------
640 PRINT
650 PRINT
660 PRINT
670 PRINT "DO YOU WANT ME TO CONVERT ANOTHER NUMBER";
680 INPUT Q$
690 IF Q$>="Y" THEN 240
700 END
```

OUTPUT:

THIS PROGRAM CONVERTS A NUMBER FROM BASE 10 TO ANY BASE.
IF YOU ARE CONVERTING YOUR NUMBER TO A BASE GREATER THAN
10, I WILL USE THE FOLLOWING CODE IN PRINTING DIGITS
GREATER THAN 9:

: FOR 10	; FOR 11	< FOR 12	= FOR 13	> FOR 14
? FOR 15	@ FOR 16	A FOR 17	B FOR 18	C FOR 19
D FOR 20	E FOR 21	F FOR 22	G FOR 23	H FOR 24
I FOR 25	J FOR 26	K FOR 27	L FOR 28	M FOR 29
N FOR 30	O FOR 31	P FOR 32	Q FOR 33	R FOR 34
S FOR 35	T FOR 36	U FOR 37	V FOR 38	W FOR 39
X FOR 40	Y FOR 41	Z FOR 42	[FOR 43	\ FOR 44
] FOR 45	↑ FOR 46	← FOR 47		

WHAT NUMBER DO YOU WANT CONVERTED ?57
IN WHAT BASE WOULD YOU LIKE THE NUMBER ?2

   57 IN BASE 2 IS 111001

288   Computer Programming in BASIC

```
DO YOU WANT ME TO CONVERT ANOTHER NUMBER ?YES


WHAT NUMBER DO YOU WANT CONVERTED ?463.902
IN WHAT BASE WOULD YOU LIKE THE NUMBER ?47

   463.902 IN BASE 47 IS 9X.ZBHB



DO YOU WANT ME TO CONVERT ANOTHER NUMBER ?NO
```

Program 9: Logarithm of N in Base B

```
10 PRINT "THIS PROGRAM FINDS THE LOGARITHM OF A NUMBER IN ANY BASE."
20 DEF FNL(B,X)=LOG(X)/LOG(B)
30 PRINT
40 PRINT
50 PRINT "WHAT IS THE NUMBER ";
60 INPUT X
65 X=ABS(X)
70 PRINT "WHAT IS THE BASE ";
80 INPUT B
90 IF B>1 THEN 120
100 PRINT "ARE YOU KIDDING ?"
110 GOTO 160
120 IF X<>0 THEN 150
130 PRINT "LOG"B;"(0) IS NEGATIVE INFINITY."
140 GOTO 160
150 PRINT "LOG"B;"("X;")=" FNL(B,X)
160 PRINT
170 PRINT
180 PRINT "DO YOU WANT ME TO FIND ANOTHER LOGARITHM ";
190 INPUT Q$
200 IF Q$>="Y" THEN 30
210 END
```

OUTPUT:

```
THIS PROGRAM FINDS THE LOGARITHM OF A NUMBER IN ANY BASE.


WHAT IS THE NUMBER   ?123456
WHAT IS THE BASE   ?10
LOG 10 ( 123456 )= 5.09151


DO YOU WANT ME TO FIND ANOTHER LOGARITHM   ?YES


WHAT IS THE NUMBER   ?123456
WHAT IS THE BASE   ?2.71828
LOG 2.71828 ( 123456 )= 11.7236


DO YOU WANT ME TO FIND ANOTHER LOGARITHM   ?NO
```

Program 10: Approximation of the Square Root of N

```
10 PRINT "THIS PROGRAM APPROXIMATES THE SQUARE ROOT OF A NUMBER."
20 PRINT
30 PRINT
40 PRINT "WHAT IS THE NUMBER ";
50 INPUT N
60 PRINT "HOW MANY APPROXIMATIONS ";
70 INPUT R
75 REM----------CALCULATING SQUARE ROOT----------
80 N=ABS(N)
90 A=N/2      '----INITIAL GUESS FOR SQUARE ROOT
100 FOR I=1 TO R
110     A=(N/A+A)/2
120 NEXT I
130 PRINT
140 PRINT "THE SQUARE ROOT OF"N;"USING"R;"APPROXIMATIONS IS"A
150 PRINT "THE SQUARE ROOT OF"N;"USING THE BUILT-IN FUNCTION IS"SQR(N)
160 PRINT "THE TWO VALUES DIFFER BY"ABS(A-SQR(N))
170 PRINT
180 PRINT
190 PRINT
195 REM--------------REPEAT ?---------------
200 PRINT "DO YOU WANT ME TO FIND ANOTHER SQUARE ROOT ";
210 INPUT Q$
220 IF Q$="Y" THEN 20
230 END
```

OUTPUT:

```
THIS PROGRAM APPROXIMATES THE SQUARE ROOT OF A NUMBER.

WHAT IS THE NUMBER  ?2
HOW MANY APPROXIMATIONS  ?4

THE SQUARE ROOT OF 2 USING 4 APPROXIMATIONS IS 1.41421
THE SQUARE ROOT OF 2 USING THE BUILT-IN FUNCTION IS 1.41421
THE TWO VALUES DIFFER BY 0


DO YOU WANT ME TO FIND ANOTHER SQUARE ROOT  ?YES


WHAT IS THE NUMBER  ?10
HOW MANY APPROXIMATIONS  ?3

THE SQUARE ROOT OF 10 USING 3 APPROXIMATIONS IS 3.16232
THE SQUARE ROOT OF 10 USING THE BUILT-IN FUNCTION IS 3.16228
THE TWO VALUES DIFFER BY 4.17531 E-5


DO YOU WANT ME TO FIND ANOTHER SQUARE ROOT  ?YES


WHAT IS THE NUMBER  ?12345
HOW MANY APPROXIMATIONS  ?20
```

```
THE SQUARE ROOT OF 12345 USING 20 APPROXIMATIONS IS 111.108
THE SQUARE ROOT OF 12345 USING THE BUILT-IN FUNCTION IS 111.108
THE TWO VALUES DIFFER BY 0


DO YOU WANT ME TO FIND ANOTHER SQUARE ROOT ?NO
```

Program 11: To Find the Rth Root of N

```
100 PRINT "THIS PROGRAM FINDS THE RTH ROOT OF A NUMBER N."
110 PRINT
120 PRINT
130 PRINT "WHAT ARE YOUR VALUES OF R AND N ";
140 INPUT R,N
150 IF R*N=0 THEN 240
160 X=N/R
170 FOR I=1 TO 1000
180    A=X
190    X=(N/X↑(R-1)+(R-1)*X)/R
200    IF ABS(A-X)/(ABS(A)+ABS(X))<1E-6 THEN 220
210 NEXT I
220 PRINT "THE"R;"TH ROOT OF"N;"IS"X
230 GO TO 120
240 END
```

OUTPUT:

```
THIS PROGRAM FINDS THE RTH ROOT OF A NUMBER N.

WHAT ARE YOUR VALUES OF R AND N ?4,16
THE 4 TH ROOT OF 16    IS 2

WHAT ARE YOUR VALUES OF R AND N ?5.6722,685.331
THE 5.6722  TH ROOT OF 685.331 IS 3.16202

WHAT ARE YOUR VALUES OF R AND N ?1,0
```

Program 12: To Divide One Polynomial By Another

```
100 DIM A(36),D(36),N$(36)
110 PRINT "THIS PROGRAM DIVIDES ONE POLYNOMIAL BY ANOTHER.  WHEN"
120 PRINT "THE COMPUTER ASKS YOU, GIVE THE COEFFICIENTS OF THE"
```

```
130 PRINT "POLYNOMIAL BEING DIVIDED AND THEN THE COEFFICIENTS OF"
140 PRINT "THE DIVISOR."
150 MAT READ N$
160 PRINT
170 PRINT
180 PRINT
190 REM-----------ENTERING POLYNOMIALS-------------
200 PRINT "DIVIDEND";
210 MAT INPUT A
220 N1=NUM
230 IF N1<=36 THEN 260
240 PRINT "SORRY.  I CAN'T HANDLE POLYNOMIALS THAT BIG."
250 GOTO 550
260 PRINT "DIVISOR";
270 MAT INPUT D
280 N2=NUM
290 IF N2>36 THEN 240
300 PRINT
310 REM-----------PRINTING POLYNOMIALS-------------
320 FOR I=1 TO N1-1
330     PRINT A(I);N$(N1-I);"+";
340 NEXT I
350 PRINT A(N1);"DIVIDED BY";
360 FOR I=1 TO N2-1
370     PRINT D(I);N$(N2-I);"+";
380 NEXT I
390 PRINT D(N2);"EQUALS: "
400 REM----------DIVIDING POLYNOMIALS-------------
410 FOR K=1 TO N1-N2+1
420     PRINT A(K)/D(1);N$(N1-N2+1-K);
430     IF K=N1-N2+1 THEN 450
440     PRINT "+";
450     FOR H=2 TO N2
460         A(H+K-1)=A(H+K-1)-D(H)*A(K)/D(1)
470     NEXT H
480 NEXT K
490 PRINT
500 REM----------REMAINDER------------
510 FOR H=N1-N2+2 TO N1
520     IF A(H)<>0 THEN 630
530 NEXT H
540 REM----------REPEAT ?------------
550 PRINT
560 PRINT
570 PRINT
580 PRINT "DO YOU WISH TO RUN AGAIN";
590 INPUT Q$
600 IF Q$>="Y" THEN 160
610 STOP
620 REM----------PRINTING REMAINDER------------
630 PRINT "REMAINDER: ";
640 FOR K=H TO N1-1
650     PRINT A(K);N$(N2-I);"+";
660 NEXT K
670 PRINT A(N1)
680 GOTO 550
690 DATA X,X↑2,X↑3,X↑4,X↑5,X↑6,X↑7,X↑8,X↑9,X↑10,X↑11,X↑12,X↑13,X↑14
700 DATA X↑15,X↑16,X↑17,X↑18,X↑19,X↑20,X↑21,X↑22,X↑23,X↑24,X↑25
710 DATA X↑26,X↑27,X↑28,X↑29,X↑30,X↑31,X↑32,X↑33,X↑34,X↑35,X↑36
720 END
```

OUTPUT:

```
THIS PROGRAM DIVIDES ONE POLYNOMIAL BY ANOTHER.  WHEN
THE COMPUTER ASKS YOU, GIVE THE COEFFICIENTS OF THE
POLYNOMIAL BEING DIVIDED AND THEN THE COEFFICIENTS OF
THE DIVISOR.


DIVIDEND ?1,0,0,1
DIVISOR ?1,1

 1 X↑3+ 0 X↑2+ 0 X+ 1 DIVIDED BY 1 X+ 1 EQUALS:
 1 X↑2+-1 X+ 1


DO YOU WISH TO RUN AGAIN ?YES


DIVIDEND ?1,2,3,4,5
DIVISOR ?2,3,4

 1 X↑4+ 2 X↑3+ 3 X↑2+ 4 X+ 5 DIVIDED BY 2 X↑2+ 3 X+ 4 EQUALS:
 0.5 X↑2+ 0.25 X+ 0.125
REMAINDER: 2.625 X+ 4.5


DO YOU WISH TO RUN AGAIN ?NO
```

Program 13: To Graph a Function With X-Axis Horizontal

```
100 DEF FNA(X)=EXP(-X↑2/2)/SQR(2*3.14159)      'FUNCTION TO BE PLOTTED
110 DEF FNX(X)=INT((X-A)/R+1.5)
120 DEF FNY(X)=INT((X-C)/L+1.5)
130 DIM P(50,60)
140 PRINT "THIS PROGRAM ILLUSTRATES A METHOD OF PLOTTING A FUNCTION"
150 PRINT "WITH THE X-AXIS HORIZONTAL AND THE Y-AXIS VERTICAL.  THE"
160 PRINT "TECHNIQUE USED IS TO FILL A MATRIX WITH THE POINTS OF THE"
170 PRINT "FUNCTION IN THE PORTION PLOTTED."
180 PRINT
190 PRINT
200 PRINT
210 READ A,B,C,D,L      '---DOMAIN, RANGE, INCREMENTS ALONG Y-AXIS
220 R=(B-A)/55
230 REM--SETTING UP MATRIX WITH ONES WHERE POINTS SHOULD BE PLOTTED--
240 MAT P=ZER(55,FNY(D))
250 C1=C-L/2
260 D1=D+L/2
270 FOR X=A TO B STEP R
280     IF FNA(X)>=D1 THEN 310
290     IF FNA(X)<C1 THEN 310
300     P(FNX(X),FNY(FNA(X)))=1    '---FLAG FOR POINT TO BE PLOTTED
310 NEXT X
```

Appendix A: Algebra Programs   293

```
320 REM------PRINTING POINTS STORED IN MATRIX------
330 FOR Y=D TO C STEP -L
340     PRINT Y;TAB(14);":";
350     FOR X=1 TO 55
360         IF P(X,FNY(Y))=0 THEN 380
370         PRINT TAB(X+15);".";
380     NEXT X
390     PRINT
400 NEXT Y
410 REM-----------X-AXIS------------
420 PRINT TAB(14);"+";
430 PRINT "-----------------------------------------------------------"
440 PRINT TAB(14);"X-AXIS: FROM ";A;"TO";B;"IN INTERVALS OF";R
450 DATA -3,3, 0,0.4, 0.025
460 END
```

OUTPUT:

THIS PROGRAM ILLUSTRATES A METHOD OF PLOTTING A FUNCTION
WITH THE X-AXIS HORIZONTAL AND THE Y-AXIS VERTICAL. THE
TECHNIQUE USED IS TO FILL A MATRIX WITH THE POINTS OF THE
FUNCTION IN THE PORTION PLOTTED.

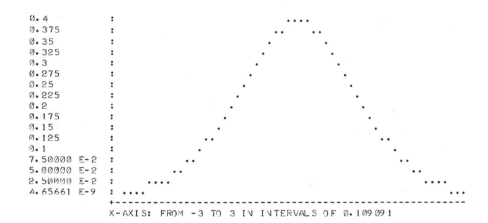

Program 14: To Graph Up to 26 Functions

```
10 DIM Y(60)
20 REM-----------DEFINING FUNCTIONS------------
30 PRINT "THIS PROGRAM PLOTS UP TO 26 FUNCTIONS."
40 PRINT "HAVE YOU DEFINED YOUR FUNCTIONS";
50 INPUT Q$
60 IF Q$>="Y" THEN 170
```

```
70 PRINT "TYPE: 260 FNZ=(FIRST FUNCTION)"
80 PRINT "      280 FNZ=(SECOND FUNCTION)"
90 PRINT "      300 FNZ=(THIRD FUNCTION), ETC."
100 PRINT "[X IS THE INDEPENDENT VARIABLE OF YOUR FUNCTIONS.]"
110 PRINT "THEN TYPE 'RUN' <RETURN>."
120 IF N>0 THEN 160
130 PRINT
140 PRINT "I WILL TYPE 'A' FOR A POINT ON THE GRAPH OF THE 1ST"
150 PRINT "FUNCTION, 'B' FOR A POINT OF THE 2ND FUNCTION, ETC."
160 STOP
170 PRINT "HOW MANY FUNCTIONS DO YOU WANT PLOTTED";
180 INPUT N
190 N=INT(N)
200 IF N<1 THEN 170
210 IF N>26 THEN 30
220 DEF FNZ(X,Y)
230     IF Y>13 THEN 250
240     ON Y GOTO 260,280,300,320,340,360,380,400,420,440,460,480,500
250     ON Y-13 GOTO 520,540,560,580,600,620,640,660,680,700,720,740,760
260     FNZ=0
270     GO TO 770
280     FNZ=0
290     GO TO 770
300     FNZ=0
310     GO TO 770
320     FNZ=0
330     GO TO 770
340     FNZ=0
350     GO TO 770
360     FNZ=0
370     GO TO 770
380     FNZ=0
390     GO TO 770
400     FNZ=0
410     GO TO 770
420     FNZ=0
430     GO TO 770
440     FNZ=0
450     GO TO 770
460     FNZ=0
470     GO TO 770
480     FNZ=0
490     GO TO 770
500     FNZ=0
510     GO TO 770
520     FNZ=0
530     GO TO 770
540     FNZ=0
550     GO TO 770
560     FNZ=0
570     GO TO 770
580     FNZ=0
590     GO TO 770
600     FNZ=0
610     GO TO 770
620     FNZ=0
630     GO TO 770
640     FNZ=0
650     GO TO 770
660     FNZ=0
670     GO TO 770
680     FNZ=0
690     GO TO 770
700     FNZ=0
```

```
710     GOTO 770
720     FNZ=0
730     GOTO 770
740     FNZ=0
750     GOTO 770
760     FNZ=0
770 FNEND
780 PRINT
790 PRINT
800 REM------------DOMAIN AND RANGE------------
810 PRINT "WHAT ARE THE LOWER AND UPPER BOUNDS ALONG THE X-AXIS";
820 INPUT A,B
830 C=FNZ(A,1)
840 D=FNZ(A,1)
850 FOR S=A TO B STEP (B-A)/100
860     FOR T=1 TO N
870         IF C<=FNZ(S,T) THEN 890 '----MINIMUM
880         C=FNZ(S,T)
890         IF D>=FNZ(S,T) THEN 910 '----MAXIMUM
900         D=FNZ(S,T)
910     NEXT T
920 NEXT S
930 PRINT "I SUGGEST A RANGE FROM"C"TO"D
940 PRINT "WHAT ARE THE LOWER AND UPPER BOUNDS ALONG THE Y-AXIS";
950 INPUT C,D
960 L=(D-C)/55
970 R=5/3*L
980 PRINT "WHAT IS YOUR INCREMENT ALONG THE X-AXIS";
990 PRINT " (I SUGGEST"R")";
1000 INPUT R
1010 REM----------LABELING Y-AXIS----------
1020 PRINT
1030 PRINT
1040 PRINT
1050 PRINT
1060 PRINT TAB(14);"Y-AXIS: FROM"C"TO"D;"IN INTERVALS OF"L
1070 IF D-C<13 THEN 1130
1080 PRINT TAB(14);"+";
1090 FOR Q=1 TO 55
1100     PRINT "-";
1110 NEXT Q
1120 GOTO 1260
1130 FOR Y=C TO D STEP L
1140     IF ABS(Y-INT(Y+0.5))>=L/2 THEN 1160
1150     PRINT TAB((Y-C)/L+14.5);INT(Y+0.5);
1160 NEXT Y
1170 PRINT
1180 PRINT TAB(15);
1190 FOR Y=C TO D STEP L
1200     IF ABS(Y-INT(Y+0.5))>=L/2 THEN 1230
1210     PRINT "+";
1220     GOTO 1240
1230     PRINT "-";
1240 NEXT Y
1250 PRINT
1260 REM----------PLOTTING FUNCTIONS----------
1270 G(0)=1
1280 C1=C-L/2
1290 D1=D+L/2
1300 FOR S=A TO B STEP R
1310     FOR I=1 TO 55
1320         Y(I)=32       '----CODE NUMBER FOR SPACE
1330     NEXT I
1340     IF C>=0 THEN 1360
```

```
1350     Y(INT(1.49999-C/L))=58      '----CODE NUMBER FOR COLON
1360     FOR I=N TO 1 STEP -1
1370         IF FNZ(S,I)>=D1 THEN 1400
1380         IF FNZ(S,I)<C1 THEN 1400
1390         Y(INT((FNZ(S,I)-C)/L+1.5))=I+64    'CODE NO. FOR LETTER
1400     NEXT I
1410     REM----ELIMINATING SPACES AT END OF STRING----
1420     Y(0)=0
1430     FOR I=55 TO 1 STEP -1
1440         IF Y(I)<>32 THEN 1460
1450         IF Y(0)=0 THEN 1470
1460         Y(0)=Y(0)+1
1470     NEXT I
1480     CHANGE Y TO Y$
1490     PRINT S;TAB(14);":";Y$
1500 NEXT S
1510 REM---------REPEAT ?----------
1520 PRINT
1530 PRINT
1540 PRINT
1550 PRINT
1560 PRINT "DO YOU WANT ME TO PLOT ANOTHER SET OF FUNCTIONS";
1570 INPUT R$
1580 IF R$>="Y" THEN 40
1590 END
```

OUTPUT:

```
THIS PROGRAM PLOTS UP TO 26 FUNCTIONS.
HAVE YOU DEFINED YOUR FUNCTIONS ?NO
TYPE: 260 FNZ=(FIRST FUNCTION)
      280 FNZ=(SECOND FUNCTION)
      300 FNZ=(THIRD FUNCTION), ETC.
[X IS THE INDEPENDENT VARIABLE OF YOUR FUNCTIONS.]
THEN TYPE 'RUN' <RETURN>.

I WILL TYPE 'A' FOR A POINT ON THE GRAPH OF THE 1ST
FUNCTION, 'B' FOR A POINT OF THE 2ND FUNCTION, ETC.




260 FNZ=EXP(-X/2)*COS(3*X+3.14159/4)
280 FNZ=EXP(-X/2)
300 FNZ=-EXP(-X/2)


    RUN



THIS PROGRAM PLOTS UP TO 26 FUNCTIONS.
HAVE YOU DEFINED YOUR FUNCTIONS ?YES
HOW MANY FUNCTIONS DO YOU WANT PLOTTED ?3

WHAT ARE THE LOWER AND UPPER BOUNDS ALONG THE X-AXIS ?-1,3
I SUGGEST A RANGE FROM-1.64872 TO 1.64872
WHAT ARE THE LOWER AND UPPER BOUNDS ALONG THE Y-AXIS ?-1.65,1.65
WHAT IS YOUR INCREMENT ALONG THE X-AXIS (I SUGGEST 0.1 ) ?0.1
```

Appendix A: Algebra Programs   297

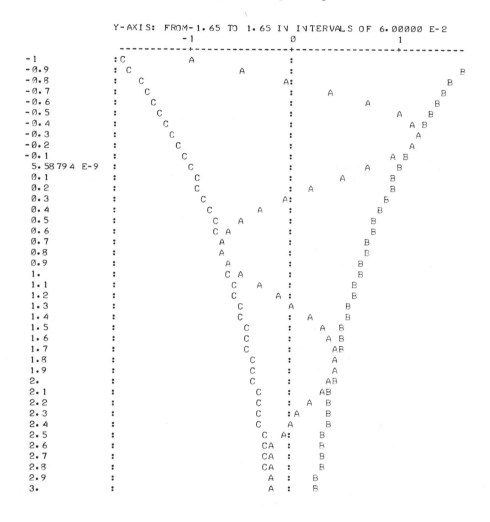

DO YOU WANT ME TO PLOT ANOTHER SET OF FUNCTIONS ?NO

Program 15: Real Zeros of any Function by Newton's Method

```
10 DEF FNC(X)=X
20 DIM R(20)
30 PRINT "THIS PROGRAM FINDS THE REAL ZEROS OF ANY FUNCTION."
40 PRINT
50 PRINT "HAVE YOU DEFINED YOUR FUNCTION ";
60 INPUT Q$
70 IF Q$>="Y" THEN 110
```

```
80 PRINT "TYPE '10 DEF FNC(X)=(YOUR FUNCTION)',<RETURN>,'RUN'."
90 STOP
100 REM----------INTERVAL----------
110 PRINT
120 PRINT "WHAT IS THE LOWER LIMIT OF YOUR INTERVAL ";
130 INPUT A1
140 PRINT "WHAT IS THE UPPER LIMIT OF YOUR INTERVAL ";
150 INPUT B
160 PRINT
170 PRINT
180 LET A=A1
190 LET N=-1
200 REM----------SEARCH FOR CHANGE OF SIGN----------
210 LET N=N+1
220 LET V=FNC(A)
230 LET H=(B-A)/100
240 FOR X=A TO B STEP H
250     LET S=FNC(X)
260     IF S=0 THEN 380
270     IF V*S<0 THEN 330
280 NEXT X
290 IF N>0 THEN 420
300 PRINT "THERE ARE NO REAL ZEROS IN YOUR INTERVAL."
310 GOTO 480
320 REM----------NEWTON'S METHOD----------
330 LET X=X-H/2
340 LET X=X+FNC(X)/(FNC(X)-FNC(X+1E-6))*1E-6
350 LET X1=X
360 LET X=X1+FNC(X1)/(FNC(X1)-FNC(X1+1E-6))*1E-6
370 IF ABS(X1-X)/(ABS(X1)+ABS(X))>=1E-7 THEN 350
380 LET R(N)=X
390 LET A=X+1E-5
400 IF A<B THEN 210
410 REM----------PRINTOUT----------
420 PRINT "THE REAL ZEROS FROM"A1;"TO"B;"ARE:"
430 FOR I=0 TO N-1
440     PRINT R(I);
450 NEXT I
460 REM----------REPEAT?----------
470 PRINT
480 PRINT
490 PRINT
500 PRINT "DO YOU WISH TO RUN AGAIN ";
510 INPUT Q$
520 IF Q$>="Y" THEN 40
530 END
```

OUTPUT:

```
THIS PROGRAM FINDS THE REAL ZEROS OF ANY FUNCTION.

HAVE YOU DEFINED YOUR FUNCTION ?NO
TYPE '10 DEF FNC[X]=[YOUR FUNCTION]',<RETURN>,'RUN'.



10 DEF FNC(X)=X↑2-2*X-3
RUN
```

THIS PROGRAM FINDS THE REAL ZEROS OF ANY FUNCTION.

HAVE YOU DEFINED YOUR FUNCTION ?YES

WHAT IS THE LOWER LIMIT OF YOUR INTERVAL ?-5
WHAT IS THE UPPER LIMIT OF YOUR INTERVAL ?5

THE REAL ZEROS FROM-5 TO 5 ARE:
-1 3.

DO YOU WISH TO RUN AGAIN ?YES

HAVE YOU DEFINED YOUR FUNCTION ?NO
TYPE '10 DEF FNC[X]=[YOUR FUNCTION]',<RETURN>,'RUN'.

10 DEF FNC(X)=X↑2+1
RUN

THIS PROGRAM FINDS THE REAL ZEROS OF ANY FUNCTION.

HAVE YOU DEFINED YOUR FUNCTION ?YES

WHAT IS THE LOWER LIMIT OF YOUR INTERVAL ?-10
WHAT IS THE UPPER LIMIT OF YOUR INTERVAL ?10

THERE ARE NO REAL ZEROS IN YOUR INTERVAL.

DO YOU WISH TO RUN AGAIN ?YES

HAVE YOU DEFINED YOUR FUNCTION ?NO
TYPE '10 DEF FNC[X]=[YOUR FUNCTION]',<RETURN>,'RUN'.

10 DEF FNC(X)=SIN(X)-COS(X)
RUN

THIS PROGRAM FINDS THE REAL ZEROS OF ANY FUNCTION.

HAVE YOU DEFINED YOUR FUNCTION ?YES

WHAT IS THE LOWER LIMIT OF YOUR INTERVAL ?-10
WHAT IS THE UPPER LIMIT OF YOUR INTERVAL ?10

THE REAL ZEROS FROM-10   TO 10   ARE:
-8.63938 -5.49779 -2.35619 .785398 3.92699 7.06858

DO YOU WISH TO RUN AGAIN ?NO

APPENDIX B
Geometry Programs

Program 16: Pythagorean Triplets

```
10 PRINT "THIS PROGRAM FINDS PYTHAGOREAN TRIPLETS."
20 PRINT "WHAT IS THE UPPER LIMIT OF THE TRIPLETS";
30 INPUT N
40 PRINT
50 PRINT
60 IF N>4 THEN 90
70 PRINT "THE TRIPLETS BEGIN ABOVE 4."
80 STOP
90 PRINT "LEG", "LEG", "HYPOTENUSE"
100 PRINT
110 FOR A=3 TO N
120     FOR B=A TO N
130         FOR C=5 TO N
140             IF A↑2+B↑2<>C↑2 THEN 160
150             PRINT A,B,C
160         NEXT C
170     NEXT B
180 NEXT A
190 END
```

OUTPUT:

```
THIS PROGRAM FINDS PYTHAGOREAN TRIPLETS.
WHAT IS THE UPPER LIMIT OF THE TRIPLETS? 30
```

LEG	LEG	HYPOTENUSE
3	4	5
5	12	13
6	8	10
7	24	25
8	15	17
9	12	15
10	24	26
12	16	20
15	20	25
18	24	30
20	21	29

Program 17: Area of any Regular Polygon

```
100 PRINT "THIS PROGRAM FINDS THE AREA OF A REGULAR POLYGON, GIVEN"
110 PRINT "THE NUMBER OF SIDES AND EITHER THE LENGTH OF EACH SIDE"
120 PRINT "OR THE RADIUS OF THE CIRCUMSCRIBED CIRCLE."
130 PRINT
140 PRINT
150 PRINT "HOW MANY SIDES DOES THE POLYGON HAVE";
160 INPUT N
170 PRINT "ARE YOU GIVEN THE LENGTH OF EACH SIDE OR THE RADIUS OF"
180 PRINT "THE CIRCUMSCRIBED CIRCLE";
190 INPUT C$
200 IF C$>="R" THEN 250
210 PRINT "WHAT IS THE LENGTH OF THE SIDES";
220 INPUT L
230 A=N*L↑2*COT(3.14159/N)/4
240 GO TO 280
250 PRINT "WHAT IS THE RADIUS OF THE CIRCUMSCRIBED CIRCLE";
260 INPUT R
270 A=N*R↑2*SIN(2*3.14159/N)/2
280 PRINT
290 PRINT "THE AREA OF THE POLYGON IS"A
300 PRINT
310 PRINT
320 PRINT "DO YOU WANT ME TO FIND THE AREA OF ANOTHER POLYGON";
330 INPUT Q$
340 IF Q$>="Y" THEN 130
350 END
```

OUTPUT:

```
THIS PROGRAM FINDS THE AREA OF A REGULAR POLYGON, GIVEN
THE NUMBER OF SIDES AND EITHER THE LENGTH OF EACH SIDE
OR THE RADIUS OF THE CIRCUMSCRIBED CIRCLE.


HOW MANY SIDES DOES THE POLYGON HAVE ?3
ARE YOU GIVEN THE LENGTH OF EACH SIDE OR THE RADIUS OF
THE CIRCUMSCRIBED CIRCLE ?LENGTH
WHAT IS THE LENGTH OF THE SIDES ?2

THE AREA OF THE POLYGON IS 1.73205


DO YOU WANT ME TO FIND THE AREA OF ANOTHER POLYGON ?YES


HOW MANY SIDES DOES THE POLYGON HAVE ?3
ARE YOU GIVEN THE LENGTH OF EACH SIDE OR THE RADIUS OF
THE CIRCUMSCRIBED CIRCLE ?RADIUS
WHAT IS THE RADIUS OF THE CIRCUMSCRIBED CIRCLE ?1.1547

THE AREA OF THE POLYGON IS 1.73205


DO YOU WANT ME TO FIND THE AREA OF ANOTHER POLYGON ?YES
```

```
HOW MANY SIDES DOES THE POLYGON HAVE ?13
ARE YOU GIVEN THE LENGTH OF EACH SIDE OR THE RADIUS OF
THE CIRCUMSCRIBED CIRCLE ?LENGTH
WHAT IS THE LENGTH OF THE SIDES ?1.7928

THE AREA OF THE POLYGON IS 42.3808


DO YOU WANT ME TO FIND THE AREA OF ANOTHER POLYGON ?YES


HOW MANY SIDES DOES THE POLYGON HAVE ?29
ARE YOU GIVEN THE LENGTH OF EACH SIDE OR THE RADIUS OF
THE CIRCUMSCRIBED CIRCLE ?RADIUS
WHAT IS THE RADIUS OF THE CIRCUMSCRIBED CIRCLE ?3.16298

THE AREA OF THE POLYGON IS 31.1845


DO YOU WANT ME TO FIND THE AREA OF ANOTHER POLYGON ?NO
```

Program 18: To Find the Interior Angle, Diagonals of A Regular Polygon

```
100 PRINT "THIS PROGRAM DETERMINES THE MEASURE OF EACH ANGLE AND"
110 PRINT "THE NUMBER OF DIAGONALS IN A REGULAR POLYGON, GIVEN"
120 PRINT "THE NUMBER OF SIDES."
130 PRINT
140 PRINT
150 PRINT "HOW MANY SIDES";
160 INPUT N
170 PRINT
180 N=INT(N)
190 IF N<3 THEN 150
200 PRINT "MEASURE OF INTERIOR ANGLE =";(N-2)/N*180;"DEGREES."
210 PRINT "NUMBER OF DIAGONALS =";N*(N-3)/2
220 PRINT
230 PRINT
240 PRINT "DO YOU WISH TO RUN AGAIN";
250 INPUT Q$
260 IF Q$>="Y" THEN 130
270 END
```

OUTPUT:

```
THIS PROGRAM DETERMINES THE MEASURE OF EACH ANGLE AND
THE NUMBER OF DIAGONALS IN A REGULAR POLYGON, GIVEN
THE NUMBER OF SIDES.


HOW MANY SIDES ?3

MEASURE OF INTERIOR ANGLE = 60 DEGREES.
NUMBER OF DIAGONALS = 0
```

```
DO YOU WISH TO RUN AGAIN ?YES

HOW MANY SIDES ?5

MEASURE OF INTERIOR ANGLE = 108 DEGREES.
NUMBER OF DIAGONALS = 5

DO YOU WISH TO RUN AGAIN ?YES

HOW MANY SIDES ?10001

MEASURE OF INTERIOR ANGLE = 179.964 DEGREES.
NUMBER OF DIAGONALS = 49994999

DO YOU WISH TO RUN AGAIN ?NO
```

APPENDIX C
Trigonometry Programs

Program 19: Area of Triangle, Given SAS or SSS

```
100 PRINT "THIS PROGRAM FINDS THE AREA OF A TRIANGLE GIVEN EITHER"
110 PRINT "TWO SIDES AND THE INCLUDED ANGLE OR THREE SIDES."
120 PRINT
130 PRINT
140 PRINT
150 PRINT "ARE YOU GIVEN SAS OR SSS";
160 INPUT SS
170 REM----------ENTERING LENGTHS OF TWO SIDES----------
180 PRINT
190 PRINT "WHAT IS THE LENGTH OF THE FIRST SIDE ";
200 INPUT A
210 PRINT "WHAT IS THE LENGTH OF THE SECOND SIDE";
220 INPUT B
230 IF SS>="SS"THEN 300
240 REM---------------SAS---------------
250 PRINT "WHAT IS THE MEASURE OF THE INCLUDED ANGLE (IN DEGREES)";
260 INPUT C
270 R=A*B*SIN(C*3.14159/180)/2
280 GOTO 350
290 REM---------------SSS---------------
300 PRINT "WHAT IS THE LENGTH OF THE THIRD SIDE ";
310 INPUT C
320 S=(A+B+C)/2
330 R=SQR(S*(S-A)*(S-B)*(S-C))
340 REM----------PRINT-OUT----------
350 PRINT
360 PRINT
370 PRINT
380 PRINT "THE AREA IS"R
390 REM----------REPEAT ?----------
400 PRINT
410 PRINT "DO YOU WANT ME TO FIND THE AREA OF ANOTHER TRIANGLE";
420 INPUT RR
430 IF RR>="Y" THEN 120
440 END
```

OUTPUT:

```
THIS PROGRAM FINDS THE AREA OF A TRIANGLE GIVEN EITHER
TWO SIDES AND THE INCLUDED ANGLE OR THREE SIDES.
```

```
ARE YOU GIVEN SAS OR SSS?SAS

WHAT IS THE LENGTH OF THE FIRST SIDE ?4
WHAT IS THE LENGTH OF THE SECOND SIDE?5
WHAT IS THE MEASURE OF THE INCLUDED ANGLE [IN DEGREES]?30


THE AREA IS 5.

DO YOU WANT ME TO FIND THE AREA OF ANOTHER TRIANGLE?YES


ARE YOU GIVEN SAS OR SSS?SSS

WHAT IS THE LENGTH OF THE FIRST SIDE ?3
WHAT IS THE LENGTH OF THE SECOND SIDE?4
WHAT IS THE LENGTH OF THE THIRD SIDE ?5


THE AREA IS 6

DO YOU WANT ME TO FIND THE AREA OF ANOTHER TRIANGLE?YES


ARE YOU GIVEN SAS OR SSS?SAS

WHAT IS THE LENGTH OF THE FIRST SIDE ?36.7721
WHAT IS THE LENGTH OF THE SECOND SIDE?54.8279
WHAT IS THE MEASURE OF THE INCLUDED ANGLE [IN DEGREES]?41.6572


THE AREA IS 670.035

DO YOU WANT ME TO FIND THE AREA OF ANOTHER TRIANGLE?NO
```

APPENDIX D
Analytic Geometry Programs

Program 20: Equation of Parabola Through Three Points

```
100 PRINT "THIS PROGRAM FINDS THE EQUATION OF A PARABOLA PASSING"
110 PRINT "THROUGH THREE POINTS.  WHEN THE COMPUTER ASKS YOU, GIVE"
120 PRINT "THE X AND Y COORDINATES OF EACH POINT."
130 PRINT
140 PRINT
150 PRINT
160 PRINT "FIRST POINT";
170 INPUT X1,Y1
180 PRINT "SECOND POINT";
190 INPUT X2,Y2
200 PRINT "THIRD POINT";
210 INPUT X3,Y3
220 PRINT
230 D=X2*X1↑2+X1*X3↑2+X3*X2↑2-X3*X1↑2-X1*X2↑2-X2*X3↑2
240 IF D=0 THEN 330
250 A=(Y1*(X2-X3)+Y2*(X3-X1)+Y3*(X1-X2))/D
260 B=(Y1*(X3↑2-X2↑2)+Y2*(X1↑2-X3↑2)+Y3*(X2↑2-X1↑2))/D
270 F1=X3*X2↑2-X2*X3↑2
280 F2=X1*X3↑2-X3*X1↑2
290 F3=X2*X1↑2-X1*X2↑2
300 C=(Y1*F1+Y2*F2+Y3*F3)/D
310 PRINT "THE EQUATION IS Y=";A;"X↑2+";B;"X+";C
320 GOTO 410
330 IF X1<>X2 THEN 380
340 IF X2<>X3 THEN 380
350 PRINT "THE EQUATION FOR THESE POINTS IS NOT A FUNCTION.  IT"
360 PRINT "HAS THE FORM X=";X1
370 GOTO 410
380 PRINT "THE EQUATION FOR THESE POINTS IS NOT A FUNCTION AND THE"
390 PRINT "COEFFICIENTS CANNOT BE DETERMINED BY THIS PROGRAM."
400 GOTO 410
410 PRINT
420 PRINT
430 PRINT
440 PRINT "DO YOU WISH TO RUN AGAIN";
450 INPUT Q$
460 IF Q$>="Y" THEN 130
470 END
```

OUTPUT:

THIS PROGRAM FINDS THE EQUATION OF A PARABOLA PASSING THROUGH THREE POINTS. WHEN THE COMPUTER ASKS YOU, GIVE THE X AND Y COORDINATES OF EACH POINT.

FIRST POINT ?0,0
SECOND POINT ?1,1
THIRD POINT ?2,4

THE EQUATION IS Y= 1 X↑2+ 0 X+ 0

DO YOU WISH TO RUN AGAIN ?YES

FIRST POINT ?1,2
SECOND POINT ?3,4
THIRD POINT ?5,1

THE EQUATION IS Y=-0.625 X↑2+ 3.5 X+-0.875

DO YOU WISH TO RUN AGAIN ?YES

FIRST POINT ?1,1
SECOND POINT ?1,2
THIRD POINT ?1,3

THE EQUATION FOR THESE POINTS IS NOT A FUNCTION. IT HAS THE FORM X= 1

DO YOU WISH TO RUN AGAIN ?YES

FIRST POINT ?0,0
SECOND POINT ?2,2
THIRD POINT ?2,-2

THE EQUATION FOR THESE POINTS IS NOT A FUNCTION AND THE COEFFICIENTS CANNOT BE DETERMINED BY THIS PROGRAM.

DO YOU WISH TO RUN AGAIN ?NO

Program 21: Analysis of $AX^2+BXY+CY^2+DX+EY+F=0$

```
100 REM                     DAVID GOLUB--QUADRATIC ANALYZER
110 PRINT "THIS PROGRAM ANALYZES A QUADRATIC EQUATION IN X AND Y."
120 PRINT "THE EQUATION IS:   AX↑2+BXY+CY↑2+DX+EY+F=0  ."
130 PRINT
140 PRINT "TYPE YOUR COEFFICIENTS IN ORDER: A, B, C, D, E, F,"
150 PRINT "SEPARATED BY COMMAS."
160 PRINT
170 PRINT "WHAT IS YOUR EQUATION";
180 INPUT A, B, C, D, E, F
190 PRINT
200 IF A<>0 THEN 320
210 IF B<>0 THEN 320
220 IF C<>0 THEN 320
230 REM                    NOT QUADRATIC
240 IF D<>0 THEN 290
250 IF E<>0 THEN 290
260 PRINT "YOU HAVE GIVEN AN EQUATION WITH ALL COEFFICIENTS = 0."
270 PRINT "I CAN'T ACCEPT THAT. TRY AGAIN."
280 GOTO 1730
290 PRINT "THE EQUATION IS THE LINE:"
300 PRINT D;"X+";E;"Y=";-F
310 GOTO 1730
320 P1=3.14159
330 Q=B↑2-4*A*C
340 IF Q=0 THEN 1320
350 REM                   ELLIPSE OR HYPERBOLA
360 X=(2*C*D-B*E)/Q
370 Y=(2*A*E-B*D)/Q
380 F1=-(D*X/2+E*Y/2+F)
390 G=0
400 IF B=0 THEN 440
410 G=P1/4
420 IF A=C THEN 440
430 G=(1/2)*ATN(B/(A-C))
440 U=COS(G)
450 V=SIN(G)
460 K=A*U↑2+B*U*V+C*V↑2
470 L=A*V↑2-B*U*V+C*U↑2
480 IF K>=0 THEN 530
490 K=-K
500 L=-L
510 F1=-F1
520 REM                   NEW EQUATION: KX↑2+LY↑2=F1
530 IF Q>0 THEN 940
540 IF F1>0 THEN 610
550 IF F1=0 THEN 580
560 PRINT "THE EQUATION HAS NO REAL SOLUTION SET AND NO GRAPH."
570 GOTO 1730
580 PRINT "THE EQUATION IS A SINGLE POINT"
590 PRINT "AT (";X;",";Y;")"
600 GOTO 1730
610 IF K<>L THEN 670
620 PRINT "THE EQUATION IS A CIRCLE WITH ECCENTRICITY 0."
630 PRINT "THE CENTER IS (";X;",";Y;")"
640 PRINT "THE RADIUS IS ";SQR(F1/K)
650 PRINT "THE AREA IS ";P1*F1/K
660 GOTO 1730
670 IF K<L THEN 710
680 G=G+P1/2
```

```
690 GOTO 440
700 REM                    ELLIPSE
710 A1=SQR(F1/ABS(K))
720 B1=SQR(F1/ABS(L))
730 C1=SQR(A1↑2-B1↑2)
740 PRINT "THE EQUATION IS AN ELLIPSE WITH ECCENTRICITY "C1/A1
750 PRINT "THE CENTER IS ("X;","Y;")"
760 PRINT "THE ANGLE FROM THE X-AXIS TO THE MAJOR AXIS"
770 PRINT "IS "G*180/P1" DEGREES."
780 PRINT "THE FOCI ARE ("X+C1*U;","Y+C1*V;")"
790 PRINT "           AND ("X-C1*U;","Y-C1*V;")"
800 PRINT "THE SUM OF THE FOCAL RADII IS "2*A1
810 PRINT "THE MAJOR AXIS HAS A LENGTH OF "2*A1
820 PRINT "THE MINOR AXIS HAS A LENGTH OF "2*B1
830 PRINT "THE FOCAL CHORD HAS A LENGTH OF "2*B1↑2/A1
840 PRINT "THE MAJOR AXIS IS THE LINE:"
850 PRINT -V;"X+"U;"Y="-V*X+U*Y
860 PRINT "THE MINOR AXIS IS THE LINE:"
870 PRINT U;"X+"V;"Y="U*X+V*Y
880 PRINT "THE DIRECTRICES ARE THE LINES:"
890 PRINT U;"X+"V;"Y="U*X+V*Y+A1↑2/C1
900 PRINT "AND"
910 PRINT U;"X+"V;"Y="U*X+V*Y-A1↑2/C1
920 PRINT "THE AREA IS "P1*A1*B1
930 GOTO 1730
940 IF F1>0 THEN 1050
950 IF F1=0 THEN 980
960 G=G+P1/2
970 GOTO 440
980 B1=SQR(ABS(K))
990 A1=SQR(ABS(L))
1000 PRINT "THE EQUATION IS TWO LINES:"
1010 PRINT B1*U+A1*V;"X+"B1*V-A1*U;"Y="(B1*U+A1*V)*X+(B1*V-A1*U)*Y
1020 PRINT "AND"
1030 PRINT B1*U-A1*V;"X+"B1*V+A1*U;"Y="(B1*U-A1*V)*X+(B1*V+A1*U)*Y
1040 GOTO 1730
1050 A1=SQR(F1/ABS(K))
1060 B1=SQR(F1/ABS(L))
1070 C1=SQR(A1↑2+B1↑2)
1080 PRINT "THE EQUATION IS AN HYPERBOLA WITH ECCENTRICITY "C1/A1
1090 PRINT "THE CENTER IS ("X;","Y;")"
1100 PRINT "THE ANGLE FROM THE X-AXIS TO THE MAJOR AXIS"
1110 PRINT "IS "G*180/P1" DEGREES."
1120 PRINT "THE FOCI ARE ("X+C1*U;","Y+C1*V;")"
1130 PRINT "           AND ("X-C1*U;","Y-C1*V;")"
1140 PRINT "THE DIFFERENCE OF THE FOCAL RADII IS "2*A1
1150 PRINT "THE MAJOR AXIS HAS A LENGTH OF "2*A1
1160 PRINT "THE MINOR AXIS HAS A LENGTH OF "2*B1
1170 PRINT "THE FOCAL CHORD HAS A LENGTH OF "2*B1↑2/A1
1180 PRINT "THE MAJOR AXIS IS THE LINE:"
1190 PRINT -V;"X+"U;"Y="-V*X+U*Y
1200 PRINT "THE MINOR AXIS IS THE LINE:"
1210 PRINT U;"X+"V;"Y="U*X+V*Y
1220 PRINT "THE DIRECTRICES ARE THE LINES:"
1230 PRINT U;"X+"V;"Y="U*X+V*Y+A1↑2/C1
1240 PRINT "AND"
1250 PRINT U;"X+"V;"Y="U*X+V*Y-A1↑2/C1
1260 PRINT "THE ASYMPTOTES ARE THE LINES:"
1270 PRINT B1*U+A1*V;"X+"B1*V-A1*U;"Y="(B1*U+A1*V)*X+(B1*V-A1*U)*Y
1280 PRINT "AND"
1290 PRINT B1*U-A1*V;"X+"B1*V+A1*U;"Y="(B1*U-A1*V)*X+(B1*V+A1*U)*Y
1300 GOTO 1730
1310 REM                    PARABOLA
```

```
1320 G=0
1330 IF A=0 THEN 1370
1340 G=P1/2
1350 IF B=0 THEN 1370
1360 G=ATN(-2*A/B)
1370 L=A+C
1380 M=D*(COS(G))+E*SIN(G)
1390 N=-D*SIN(G)+E*COS(G)
1400 IF ABS(M)<1E-5 THEN 1610
1410 K1=-N/(2*L)
1420 H1=(-F+L*K1↑2)/M
1430 C1=-L/(4*M)
1440 IF C1>=0 THEN 1460
1450 C1=-C1
1460 G=G+P1
1470 REM                    EQUATION: Y↑2=4(C1)(X)
1480 PRINT "THE EQUATION IS A PARABOLA WITH ECCENTRICITY 1."
1490 PRINT "THE CENTER IS ("H1*COS(G)-K1*SIN(G);
1500 PRINT ",";H1*SIN(G)+K1*COS(G);")"
1510 PRINT "THE ANGLE FROM THE X-AXIS TO THE AXIS OF SYMMETRY"
1520 PRINT "IS ";G*180/P1;" DEGREES."
1530 PRINT "THE FOCUS IS (";(H1+C1)*COS(G)-K1*SIN(G);
1540 PRINT ",";(H1+C1)*SIN(G)+K1*COS(G);")"
1550 PRINT "THE FOCAL CHORD HAS A LENGTH OF "4*C1
1560 PRINT "THE AXIS OF SYMMETRY IS THE LINE:"
1570 PRINT -SIN(G);"X+"COS(G);"Y=";K1
1580 PRINT "THE DIRECTRIX IS THE LINE:"
1590 PRINT COS(G);"X+"SIN(G);"Y=";H1-C1
1600 GOTO 1730
1610 S=N↑2-4*L*F
1620 IF S>=0 THEN 1650
1630 PRINT "THE EQUATION HAS NO REAL SOLUTION SET AND NO GRAPH."
1640 GOTO 1730
1650 IF S>0 THEN 1690
1660 PRINT "THE EQUATION IS THE LINE:"
1670 PRINT -SIN(G);"X+"COS(G);"Y=";-N/(2*L)
1680 GOTO 1730
1690 PRINT "THE EQUATION IS TWO PARALLEL LINES:"
1700 PRINT -SIN(G);"X+"COS(G);"Y=";(-N+SQR(S))/(2*L)
1710 PRINT "AND"
1720 PRINT -SIN(G);"X+"COS(G);"Y=";(-N-SQR(S))/(2*L)
1730 PRINT
1740 PRINT
1750 PRINT
1760 PRINT
1770 PRINT "DO YOU WANT TO RUN AGAIN";
1780 INPUT GG
1790 IF GG>="Y" THEN 130
1800 END
```

OUTPUT:

THIS PROGRAM ANALYZES A QUADRATIC EQUATION IN X AND Y.
THE EQUATION IS: AX↑2+BXY+CY↑2+DX+EY+F=0 .

TYPE YOUR COEFFICIENTS IN ORDER: A, B, C, D, E, F,
SEPARATED BY COMMAS.

WHAT IS YOUR EQUATION?1,0,1,-4,8,-16

THE EQUATION IS A CIRCLE WITH ECCENTRICITY 0.
THE CENTER IS [2 ,-4]

```
THE RADIUS IS  6
THE AREA IS   113.097




DO YOU WANT TO RUN AGAIN?YES




TYPE YOUR COEFFICIENTS IN ORDER: A, B, C, D, E, F,
SEPARATED BY COMMAS.

WHAT IS YOUR EQUATION?9, 0, 16, 0, 0, -144

THE EQUATION IS AN ELLIPSE WITH ECCENTRICITY  .661438
THE CENTER IS [ 0 , 0 ]
THE ANGLE FROM THE X-AXIS TO THE MAJOR AXIS
IS  0   DEGREES.
THE FOCI ARE [ 2.64575 , 0 ]
         AND [-2.64575 , 0 ]
THE SUM OF THE FOCAL RADII IS  8
THE MAJOR AXIS HAS A LENGTH OF  8
THE MINOR AXIS HAS A LENGTH OF  6
THE FOCAL CHORD HAS A LENGTH OF  4.5
THE MAJOR AXIS IS THE LINE:
 0 X+ 1 Y= 0
THE MINOR AXIS IS THE LINE:
 1 X+ 0 Y= 0
THE DIRECTRICES ARE THE LINES:
 1 X+ 0 Y= 6.04743
AND
 1 X+ 0 Y=-6.04743
THE AREA IS   37.6991




DO YOU WANT TO RUN AGAIN?YES




TYPE YOUR COEFFICIENTS IN ORDER: A, B, C, D, E, F,
SEPARATED BY COMMAS.

WHAT IS YOUR EQUATION?1, 0, 0, 4, -1, 4

THE EQUATION IS A PARABOLA WITH ECCENTRICITY 1.
THE CENTER IS [-2.   ,-1.11892E-08  ]
THE ANGLE FROM THE X-AXIS TO THE AXIS OF SYMMETRY
IS  90    DEGREES.
THE FOCUS IS [-2.   , .250001 ]
THE FOCAL CHORD HAS A LENGTH OF  1.00001
THE AXIS OF SYMMETRY IS THE LINE:
-1 X+ 1.32807E-06   Y= 2.
THE DIRECTRIX IS THE LINE:
 1.32807E-06    X+ 1 Y=-.250004




DO YOU WANT TO RUN AGAIN?YES
```

TYPE YOUR COEFFICIENTS IN ORDER: A, B, C, D, E, F,
SEPARATED BY COMMAS.

WHAT IS YOUR EQUATION? 0, 1, 0, 0, 0, -1

THE EQUATION IS AN HYPERBOLA WITH ECCENTRICITY   1.41421
THE CENTER IS [0 , 0]
THE ANGLE FROM THE X-AXIS TO THE MAJOR AXIS
IS   45   DEGREES.
THE FOCI ARE [1.41421 , 1.41421]
        AND [-1.41421 ,-1.41421]
THE DIFFERENCE OF THE FOCAL RADII IS   2.82843
THE MAJOR AXIS HAS A LENGTH OF   2.82843
THE MINOR AXIS HAS A LENGTH OF   2.82843
THE FOCAL CHORD HAS A LENGTH OF   2.82843
THE MAJOR AXIS IS THE LINE:
-.707106 X+ .707107 Y= 0
THE MINOR AXIS IS THE LINE:
 .707107 X+ .707106 Y= 0
THE DIRECTRICES ARE THE LINES:
 .707107 X+ .707106 Y= 1
AND
 .707107 X+ .707106 Y=-1
THE ASYMPTOTES ARE THE LINES:
 2 X+-1.32713E-06   Y= 0
AND
 1.32713E-06   X+ 2 Y= 0

DO YOU WANT TO RUN AGAIN? YES

TYPE YOUR COEFFICIENTS IN ORDER: A, B, C, D, E, F,
SEPARATED BY COMMAS.

WHAT IS YOUR EQUATION?-.94563E2,9 47.2, 14.001,-.999E-2, 5.93647E-4,.035472

THE EQUATION IS AN HYPERBOLA WITH ECCENTRICITY   1.47804
THE CENTER IS [-9.33023E-07   , 1.03606E-05   ]
THE ANGLE FROM THE X-AXIS TO THE MAJOR AXIS
IS -41.7308   DEGREES.
THE FOCI ARE [9.13585E-03   ,-8.13902E-03   ]
        AND [-9.13772E-03   , 8.15974E-03   ]
THE DIFFERENCE OF THE FOCAL RADII IS   1.65667E-02
THE MAJOR AXIS HAS A LENGTH OF   1.65667E-02
THE MINOR AXIS HAS A LENGTH OF   .018031
THE FOCAL CHORD HAS A LENGTH OF   1.96248E-02
THE MAJOR AXIS IS THE LINE:
 .665631 X+ .746281 Y= 7.11085E-06
THE MINOR AXIS IS THE LINE:
 .746281 X+-.665631 Y=-7.59263E-06
THE DIRECTRICES ARE THE LINES:
 .746281 X+-.665631 Y= 5.59668E-03
AND
 .746281 X+-.665631 Y=-5.61187E-03

```
THE ASYMPTOTES ARE THE LINES:
  1.21446E-03   X+-1.21827E-02   Y=-1.27353E-07
AND
  1.22418E-02   X+ 1.80685E-04   Y=-9.54990E-09



DO YOU WANT TO RUN AGAIN?NO
```

APPENDIX E
Calculus Programs

Program 22: To Find the Limit of Any Function

```
100 DEF FNA(X)=X
110 PRINT "THIS PROGRAM FINDS THE LIMIT OF ANY FUNCTION."
120 REM-----FUNCTION, NUMBER APPROACHED, FROM WHICH SIDE-----
130 PRINT
140 PRINT "HAVE YOU DEFINED YOUR FUNCTION";
150 INPUT FF
160 IF FF>="Y" THEN 190
170 PRINT "TYPE '100 DEF FNA(X)=(YOUR FUNCTION)',<RETURN>,'RUN'."
180 STOP
190 PRINT
200 PRINT "WHAT NUMBER IS X APPROACHING (USE A LARGE NUMBER FOR";
210 PRINT " INFINITY)";
220 INPUT A
230 IF A>=1E6 THEN 290
240 IF A<=-1E6 THEN 310
250 PRINT "DOES X APPROACH"A;"FROM A VALUE GREATER THAN"A;
260 PRINT "OR LESS THAN"A;
270 INPUT SS
280 GOTO 320
290 SS="L"
300 GOTO 320
310 SS="G"
320 PRINT
330 PRINT
340 PRINT
350 PRINT
360 REM----------FINDING LIMIT----------
370 J=A/1E4+1
380 IF SS<"J" THEN 400
390 J=-A/1E4-1
400 FOR H=1 TO 10000
410     G=S
420     S=FNA(A+J)
430     J=J/2
440     IF H=1 THEN 490
450     IF ABS(J)<A/1E7 THEN 510
460     IF ABS(G-S)/(ABS(G)+ABS(S))<1E-6 THEN 540
470     IF ABS(S)>100 THEN 600
480     IF ABS(S)<0.000001 THEN 680
490 NEXT H
500 REM----------NO LIMIT----------
510 PRINT "YOUR FUNCTION HAS NO LIMIT AS X APPROACHES"A
```

```
520 GOTO 710
530 REM----------LIMIT EXISTS----------
540 PRINT "THE LIMIT OF FNA(X) AS X APPROACHES"A;
550 IF SS>"J" THEN 580
560 PRINT "FROM THE POSITIVE SIDE IS"S
570 GOTO 710
580 PRINT "FROM THE NEGATIVE SIDE IS"S
590 GOTO 710
600 REM----------FUNCTION APPROACHES INFINITY----------
610 IF S<0 THEN 650
620 PRINT "YOUR FUNCTION APPROACHES POSITIVE INFINITY AS ";
630 PRINT "X APPROACHES"A
640 GOTO 710
650 PRINT "YOUR FUNCTION APPROACHES NEGATIVE INFINITY AS ";
660 PRINT "X APPROACHES"A
670 GOTO 710
680 S=0
690 GOTO 540
700 REM----------REPEAT ?----------
710 PRINT
720 PRINT
730 PRINT
740 PRINT
750 PRINT "DO YOU WISH TO RUN AGAIN";
760 INPUT AA
770 IF AA>="Y" THEN 130
780 END
```

OUTPUT:

THIS PROGRAM FINDS THE LIMIT OF ANY FUNCTION.

HAVE YOU DEFINED YOUR FUNCTION?NO
TYPE '100 DEF FNA(X)=[YOUR FUNCTION]',<RETURN>, 'RUN'.

100 DEF FNA(X)=SIN(X)/X
RUN

THIS PROGRAM FINDS THE LIMIT OF ANY FUNCTION.

HAVE YOU DEFINED YOUR FUNCTION?YES

WHAT NUMBER IS X APPROACHING [USE A LARGE NUMBER FOR INFINITY]?0
DOES X APPROACH 0 FROM A VALUE GREATER THAN 0 OR LESS THAN 0 ?GREATER

THE LIMIT OF FNA(X) AS X APPROACHES 0 FROM THE POSITIVE SIDE IS 1

DO YOU WISH TO RUN AGAIN?YES

HAVE YOU DEFINED YOUR FUNCTION?NO
TYPE '100 DEF FNA(X)=[YOUR FUNCTION]',<RETURN>,'RUN'.

100 DEF FNA(X)=SIN(X)
RUN

THIS PROGRAM FINDS THE LIMIT OF ANY FUNCTION.

HAVE YOU DEFINED YOUR FUNCTION?YES

WHAT NUMBER IS X APPROACHING [USE A LARGE NUMBER FOR INFINITY]?1E6

YOUR FUNCTION HAS NO LIMIT AS X APPROACHES 1000000

DO YOU WISH TO RUN AGAIN?YES

HAVE YOU DEFINED YOUR FUNCTION?NO
TYPE '100 DEF FNA(X)=[YOUR FUNCTION]',<RETURN>,'RUN'.

100 DEF FNA(X)=LOG(X)
RUN

THIS PROGRAM FINDS THE LIMIT OF ANY FUNCTION.

HAVE YOU DEFINED YOUR FUNCTION?YES

WHAT NUMBER IS X APPROACHING [USE A LARGE NUMBER FOR INFINITY]?0
DOES X APPROACH 0 FROM A VALUE GREATER THAN 0 OR LESS THAN 0 ?GREATER

YOUR FUNCTION APPROACHES NEGATIVE INFINITY AS X APPROACHES 0

DO YOU WISH TO RUN AGAIN?YES

HAVE YOU DEFINED YOUR FUNCTION?NO
TYPE '100 DEF FNA(X)=[YOUR FUNCTION]',<RETURN>,'RUN'.

Appendix E: Calculus Programs   317

```
100 DEF FNA(X)=1/X
RUN
```

THIS PROGRAM FINDS THE LIMIT OF ANY FUNCTION.

HAVE YOU DEFINED YOUR FUNCTION?YES

WHAT NUMBER IS X APPROACHING [USE A LARGE NUMBER FOR INFINITY]?0
DOES X APPROACH 0 FROM A VALUE GREATER THAN 0 OR LESS THAN 0 ?GREATER

YOUR FUNCTION APPROACHES POSITIVE INFINITY AS X APPROACHES 0

DO YOU WISH TO RUN AGAIN?YES

HAVE YOU DEFINED YOUR FUNCTION?YES

WHAT NUMBER IS X APPROACHING [USE A LARGE NUMBER FOR INFINITY]?0
DOES X APPROACH 0 FROM A VALUE GREATER THAN 0 OR LESS THAN 0 ?LESS

YOUR FUNCTION APPROACHES NEGATIVE INFINITY AS X APPROACHES 0

DO YOU WISH TO RUN AGAIN?YES

HAVE YOU DEFINED YOUR FUNCTION?NO
TYPE '100 DEF FNA[X]=[YOUR FUNCTION]',<RETURN>, 'RUN'.


```
100 DEF FNA(X)=(1+X)↑(1/X)
RUN
```

THIS PROGRAM FINDS THE LIMIT OF ANY FUNCTION.

HAVE YOU DEFINED YOUR FUNCTION?YES

WHAT NUMBER IS X APPROACHING [USE A LARGE NUMBER FOR INFINITY]?0
DOES X APPROACH 0 FROM A VALUE GREATER THAN 0 OR LESS THAN 0 ?GREATER

THE LIMIT OF FNA[X] AS X APPROACHES 0 FROM THE POSITIVE SIDE IS 2.7182

DO YOU WISH TO RUN AGAIN?YES

HAVE YOU DEFINED YOUR FUNCTION?NO
TYPE '100 DEF FNA[X]=[YOUR FUNCTION]',<RETURN>,'RUN'.

100 DEF FNA(X)=X↑(1/X)
RUN

THIS PROGRAM FINDS THE LIMIT OF ANY FUNCTION.

HAVE YOU DEFINED YOUR FUNCTION?YES

WHAT NUMBER IS X APPROACHING [USE A LARGE NUMBER FOR INFINITY]?1E6

THE LIMIT OF FNA[X] AS X APPROACHES 1000000 FROM THE NEGATIVE SIDE IS
 1.00001

DO YOU WISH TO RUN AGAIN?YES

HAVE YOU DEFINED YOUR FUNCTION?NO
TYPE '100 DEF FNA[X]=[YOUR FUNCTION]',<RETURN>,'RUN'.

100 DEF FNA(X)=(X↑2-5*X+6)/(X-3)
RUN

THIS PROGRAM FINDS THE LIMIT OF ANY FUNCTION.

HAVE YOU DEFINED YOUR FUNCTION?YES

WHAT NUMBER IS X APPROACHING [USE A LARGE NUMBER FOR INFINITY]?3
DOES X APPROACH 3 FROM A VALUE GREATER THAN 3 OR LESS THAN 3 ?GREATER

THE LIMIT OF FNA[X] AS X APPROACHES 3 FROM THE POSITIVE SIDE IS 1

DO YOU WISH TO RUN AGAIN?YES

HAVE YOU DEFINED YOUR FUNCTION?NO
TYPE '100 DEF FNA[X]=[YOUR FUNCTION]',<RETURN>,'RUN'.

Appendix E: Calculus Programs   319

```
100 DEF FNA(X)=X/SQR(X+1)
RUN
```

THIS PROGRAM FINDS THE LIMIT OF ANY FUNCTION.

HAVE YOU DEFINED YOUR FUNCTION?YES

WHAT NUMBER IS X APPROACHING [USE A LARGE NUMBER FOR INFINITY]?1E6

YOUR FUNCTION APPROACHES POSITIVE INFINITY AS X APPROACHES 1000000

DO YOU WISH TO RUN AGAIN?NO

Program 23: Area Under Curve and Average Value of any Function

```
10 DEF FNA(X)=X
20 PRINT "THIS PROGRAM FINDS THE AREA UNDER THE CURVE AND THE AVERAGE"
30 PRINT "VALUE OF A FUNCTION."
40 REM------------DEFINING FUNCTION------------
50 PRINT
60 PRINT "HAVE YOU DEFINED YOUR FUNCTION ";
70 INPUT FF
80 IF FF>="Y" THEN 140
90 PRINT "TYPE'10 DEF FNA(X)=(YOUR FUNCTION) ',<RETURN>, 'RUN '."
100 PRINT
110 PRINT "****************************************************"
120 STOP
130 REM------------INTERVALS AND SUBINTERVALS------------
140 PRINT
150 PRINT
160 PRINT "WHAT IS THE LOWER BOUND OF YOUR INTERVAL ";
170 INPUT A
180 PRINT "WHAT IS THE UPPER BOUND OF YOUR INTERVAL ";
190 INPUT B
200 IF B<A THEN 160
210 PRINT "HOW MANY SUBINTERVALS DO YOU WANT ";
220 INPUT N
230 N=2*INT(N/2)
240 PRINT
250 PRINT
260 PRINT
270 PRINT
280 REM----------AREA UNDER CURVE BY SIMPSON'S RULE----------
290 S=0
300 D=(B-A)/N
310 FOR X=A+D TO B-D STEP 2*D
```

```
320     S=S+4*FNA(X)+2*FNA(X+D)
330 NEXT X
340 A1=D/3*(FNA(A)+S-FNA(B))
350 IF A<>B THEN 380
360 M=FNA(A)
370 GO TO 400
380 M=1/(B-A)*A1
390 REM---------------PRINT-OUT---------------
400 PRINT "NUMBER OF", "AVERAGE", "AREA UNDER"
410 PRINT "SUBINTERVALS", "VALUE", "CURVE"
420 PRINT
430 PRINT N,M,A1
440 PRINT
450 PRINT
460 PRINT
470 PRINT
480 PRINT "****************************************************************"
490 PRINT
500 PRINT
510 REM---------------REPEAT ?---------------
520 PRINT "DO YOU WISH TO RUN AGAIN ";
530 INPUT QQ
540 IF QQ>="Y" THEN 50
550 END
```

OUTPUT:

THIS PROGRAM FINDS THE AREA UNDER THE CURVE AND THE AVERAGE
VALUE OF A FUNCTION.

HAVE YOU DEFINED YOUR FUNCTION ?NO
TYPE '10 DEF FNA[X]=[YOUR FUNCTION]',<RETURN>,'RUN'.

****

```
10 DEF FNA(X)=X↑2
RUN
```

THIS PROGRAM FINDS THE AREA UNDER THE CURVE AND THE AVERAGE
VALUE OF A FUNCTION.

HAVE YOU DEFINED YOUR FUNCTION ?YES

WHAT IS THE LOWER BOUND OF YOUR INTERVAL ?0
WHAT IS THE UPPER BOUND OF YOUR INTERVAL ?2
HOW MANY SUBINTERVALS DO YOU WANT ?100

NUMBER OF SUBINTERVALS	AVERAGE VALUE	AREA UNDER CURVE
100	1.33333	2.66667

****

DO YOU WISH TO RUN AGAIN ?YES

HAVE YOU DEFINED YOUR FUNCTION ?NO
TYPE '10 DEF FNA(X)=[YOUR FUNCTION]',<RETURN>, 'RUN'.

****

```
10 DEF FNA(X)=SQR(1/(2*3.14159))*EXP(-X↑2/2)
RUN
```

THIS PROGRAM FINDS THE AREA UNDER THE CURVE AND THE AVERAGE VALUE OF A FUNCTION.

HAVE YOU DEFINED YOUR FUNCTION ?YES

WHAT IS THE LOWER BOUND OF YOUR INTERVAL ?-10
WHAT IS THE UPPER BOUND OF YOUR INTERVAL ?10
HOW MANY SUBINTERVALS DO YOU WANT ?1000

NUMBER OF SUBINTERVALS	AVERAGE VALUE	AREA UNDER CURVE
1000	.05	1.

****

DO YOU WISH TO RUN AGAIN ?NO

Program 24: Volume of any Solid of Revolution

```
10 DEF FNA(X)=X
20 DEF FNC(X)
30     ON F GOTO 40,60,80
40     FNC=FNA(X)↑2
50     GOTO 90
60     FNC=FNA(X)*X
70     GOTO 90
80     FNC=FNA(X)
90 FNEND
100 PRINT "THIS PROGRAM FINDS THE VOLUME OF A SOLID."
110 PRINT
120 PRINT "IS YOUR SOLID A SOLID OF REVOLUTION ";
130 INPUT S$
140 IF S$>="Y" THEN 230
150 REM----------CROSS SECTIONAL AREA FUNCTION----------
160 GOSUB 700
170 IF Q$>="Y" THEN 200
180 PRINT "TYPE '10 DEF FNA(X)=(CROSS SECTIONAL AREA FUNCTION)',"
190 PRINT "HIT <RETURN> KEY, THEN TYPE 'RUN',<RETURN>."
200 F=3
210 GOSUB 750
220 GOTO 570
230 PRINT "IS YOUR SOLID ROTATED ABOUT THE X-AXIS OR Y-AXIS ";
240 INPUT R$
250 IF R$>="Y" THEN 370
260 REM----------ROTATED ABOUT X-AXIS----------
270 GOSUB 700
280 IF Q$>="Y" THEN 320
290 PRINT "TYPE '10 DEF FNA(X)=(FUNCTION ROTATED ABOUT THE X-AXIS)',"
300 PRINT "HIT <RETURN> KEY, THEN TYPE 'RUN', <RETURN>."
310 STOP
320 F=1
330 GOSUB 750
340 V=3.14159*V
350 GOTO 570
360 REM----------ROTATED ABOUT Y-AXIS----------
370 PRINT "DO YOU WISH TO USE THE METHOD OF CYLINDRICAL SHELLS ";
380 INPUT C$
390 GOSUB 700
400 IF C$>="Y" THEN 490
410 REM----------NOT USING CYLINDRICAL SHELL METHOD----------
420 IF Q$>="Y" THEN 320
430 PRINT "TYPE '10 DEF FNA(X)=(FUNCTION ROTATED ABOUT Y-AXIS)',"
440 PRINT "THEN TYPE <RETURN>, 'RUN', <RETURN>."
450 PRINT "TO GET FNA(X) CHANGE THE X'S AND Y'S OF YOUR ORIGINAL"
460 PRINT "FUNCTION Y=F(X) AND SOLVE FOR Y."
470 STOP
480 REM----------CYLINDRICAL SHELL METHOD----------
490 IF Q$>="Y" THEN 530
500 PRINT "TYPE '10 DEF FNA(X)=(FUNCTION ROTATED ABOUT Y-AXIS)',"
510 PRINT "THEN TYPE <RETURN>, 'RUN', <RETURN>."
520 STOP
530 F=2
540 GOSUB 750
550 V=2*3.14159*V
560 REM----------PRINTOUT----------
570 PRINT
580 PRINT
590 PRINT
```

```
600 PRINT "THE VOLUME FROM"A;"TO"B;"USING"N;"SUBINTERVALS IS"V
610 PRINT
620 PRINT
630 PRINT
640 REM----------REPEAT?----------
650 PRINT "DO YOU WISH TO RUN AGAIN ";
660 INPUT AA
670 IF AA>="Y" THEN 110
680 STOP
690 REM----------SUBROUTINE FOR DEFINING FUNCTION----------
700 PRINT
710 PRINT "HAVE YOU DEFINED YOUR FUNCTION ";
720 INPUT Q$
730 RETURN
740 REM----SUBROUTINE FOR ENTERING INTERVAL AND CALCULATING VOLUME----
750 PRINT
760 PRINT "WHAT IS THE LOWER LIMIT OF YOUR INTERVAL ";
770 INPUT A
780 PRINT "WHAT IS THE UPPER LIMIT OF YOUR INTERVAL ";
790 INPUT B
800 PRINT
810 PRINT "HOW MANY SUBINTERVALS DO YOU WANT ";
820 INPUT N
830 N=2*INT(N/2)
840 D=(B-A)/N
850 REM----------DEFINITE INTEGRAL BY SIMPSON'S RULE----------
860 S=FNC(A)-FNC(B)
870 FOR X=A+D TO B-D/2 STEP 2*D
880     S=S+4*FNC(X)+2*FNC(X+D)
890 NEXT X
900 V=D/3*S
910 RETURN
920 END
```

OUTPUT:

```
THIS PROGRAM FINDS THE VOLUME OF A SOLID.

IS YOUR SOLID A SOLID OF REVOLUTION ?YES
IS YOUR SOLID ROTATED ABOUT THE X-AXIS OR Y-AXIS ?X-AXIS

HAVE YOU DEFINED YOUR FUNCTION ?NO
TYPE '10 DEF FNA(X)=(FUNCTION ROTATED ABOUT THE X-AXIS)',
HIT <RETURN> KEY, THEN TYPE 'RUN', <RETURN>.




10 DEF FNA(X)=1+2*X-X↑2
RUN



THIS PROGRAM FINDS THE VOLUME OF A SOLID.

IS YOUR SOLID A SOLID OF REVOLUTION ?YES
IS YOUR SOLID ROTATED ABOUT THE X-AXIS OR Y-AXIS ?X-AXIS

HAVE YOU DEFINED YOUR FUNCTION ?YES
```

```
WHAT IS THE LOWER LIMIT OF YOUR INTERVAL ?0
WHAT IS THE UPPER LIMIT OF YOUR INTERVAL ?2

HOW MANY SUBINTERVALS DO YOU WANT ?500


THE VOLUME FROM 0 TO 2 USING 500 SUBINTERVALS IS 13.0118


DO YOU WISH TO RUN AGAIN ?YES

IS YOUR SOLID A SOLID OF REVOLUTION ?YES
IS YOUR SOLID ROTATED ABOUT THE X-AXIS OR Y-AXIS ?Y-AXIS

DO YOU WISH TO USE THE METHOD OF CYLINDRICAL SHELLS ?NO

HAVE YOU DEFINED YOUR FUNCTION ?NO
TYPE '10 DEF FNA(X)=(FUNCTION ROTATED ABOUT Y-AXIS)',
THEN TYPE <RETURN>, 'RUN', <RETURN>.
TO GET FNA(X) CHANGE THE X'S AND Y'S OF YOUR ORIGINAL
FUNCTION Y=F(X) AND SOLVE FOR Y.



10 DEF FNA(X)=X↑2/4
RUN



THIS PROGRAM FINDS THE VOLUME OF A SOLID.

IS YOUR SOLID A SOLID OF REVOLUTION ?YES
IS YOUR SOLID ROTATED ABOUT THE X-AXIS OR Y-AXIS ?Y

DO YOU WISH TO USE THE METHOD OF CYLINDRICAL SHELLS ?NO

HAVE YOU DEFINED YOUR FUNCTION ?YES

WHAT IS THE LOWER LIMIT OF YOUR INTERVAL ?0
WHAT IS THE UPPER LIMIT OF YOUR INTERVAL ?4

HOW MANY SUBINTERVALS DO YOU WANT ?501


THE VOLUME FROM 0 TO 4 USING 500 SUBINTERVALS IS 40.2124


DO YOU WISH TO RUN AGAIN ?YES

IS YOUR SOLID A SOLID OF REVOLUTION ?YES
IS YOUR SOLID ROTATED ABOUT THE X-AXIS OR Y-AXIS ?Y-AXIS

DO YOU WISH TO USE THE METHOD OF CYLINDRICAL SHELLS ?YES

HAVE YOU DEFINED YOUR FUNCTION ?NO
TYPE '10 DEF FNA(X)=(FUNCTION ROTATED ABOUT Y-AXIS)',
THEN TYPE <RETURN>, 'RUN', <RETURN>.
```

```
10 DEF FNA(X)=1-X↑2
RUN
```

THIS PROGRAM FINDS THE VOLUME OF A SOLID.

IS YOUR SOLID A SOLID OF REVOLUTION ?YES
IS YOUR SOLID ROTATED ABOUT THE X-AXIS OR Y-AXIS ?Y

DO YOU WISH TO USE THE METHOD OF CYLINDRICAL SHELLS ?YES

HAVE YOU DEFINED YOUR FUNCTION ?YES

WHAT IS THE LOWER LIMIT OF YOUR INTERVAL ?-1
WHAT IS THE UPPER LIMIT OF YOUR INTERVAL ?1

HOW MANY SUBINTERVALS DO YOU WANT ?500

THE VOLUME FROM-1 TO 1 USING 500 SUBINTERVALS IS-2.80178 E-8

DO YOU WISH TO RUN AGAIN ?YES

IS YOUR SOLID A SOLID OF REVOLUTION ?YES
IS YOUR SOLID ROTATED ABOUT THE X-AXIS OR Y-AXIS ?Y

DO YOU WISH TO USE THE METHOD OF CYLINDRICAL SHELLS ?YES

HAVE YOU DEFINED YOUR FUNCTION ?YES

WHAT IS THE LOWER LIMIT OF YOUR INTERVAL ?0
WHAT IS THE UPPER LIMIT OF YOUR INTERVAL ?1

HOW MANY SUBINTERVALS DO YOU WANT ?500

THE VOLUME FROM 0 TO 1 USING 500 SUBINTERVALS IS 1.57080

DO YOU WISH TO RUN AGAIN ?YES

IS YOUR SOLID A SOLID OF REVOLUTION ?NO

HAVE YOU DEFINED YOUR FUNCTION ?NO
TYPE '10 DEF FNA(X)=(CROSS SECTIONAL AREA FUNCTION)',
HIT <RETURN> KEY, THEN TYPE 'RUN', <RETURN>.


```
10 DEF FNA(X)=4*X
RUN
```

THIS PROGRAM FINDS THE VOLUME OF A SOLID.

IS YOUR SOLID A SOLID OF REVOLUTION ?NO

HAVE YOU DEFINED YOUR FUNCTION ?YES

WHAT IS THE LOWER LIMIT OF YOUR INTERVAL ?0
WHAT IS THE UPPER LIMIT OF YOUR INTERVAL ?4

HOW MANY SUBINTERVALS DO YOU WANT ?7

THE VOLUME FROM 0 TO 4 USING 6 SUBINTERVALS IS 32

DO YOU WISH TO RUN AGAIN ?YES

IS YOUR SOLID A SOLID OF REVOLUTION ?YES
IS YOUR SOLID ROTATED ABOUT THE X-AXIS OR Y-AXIS ?X

HAVE YOU DEFINED YOUR FUNCTION ?NO
TYPE '10 DEF FNA(X)=(FUNCTION ROTATED ABOUT THE X-AXIS)',
HIT <RETURN> KEY, THEN TYPE 'RUN', <RETURN>.


```
10 DEF FNA(X)=SIN(X)
RUN
```


THIS PROGRAM FINDS THE VOLUME OF A SOLID.

IS YOUR SOLID A SOLID OF REVOLUTION ?YES
IS YOUR SOLID ROTATED ABOUT THE X-AXIS OR Y-AXIS ?X

HAVE YOU DEFINED YOUR FUNCTION ?YES

WHAT IS THE LOWER LIMIT OF YOUR INTERVAL ?0
WHAT IS THE UPPER LIMIT OF YOUR INTERVAL ?3.14159

HOW MANY SUBINTERVALS DO YOU WANT ?500

THE VOLUME FROM 0 TO 3.14159 USING 500 SUBINTERVALS IS 4.93480

DO YOU WISH TO RUN AGAIN ?NO

Program 25: To Calculate Sin(X) by a Power Series

```
100 PRINT "THIS PROGRAM ILLUSTRATES THE USE OF A POWER SERIES TO"
110 PRINT "DEFINE THE SINE FUNCTION. THE FUNCTION SIN(X) IS EQUAL"
120 PRINT "TO THE POWER SERIES:"
```

Appendix E: Calculus Programs   327

```
130 PRINT
140 PRINT "     X-X↑3/3!+X↑5/5!-X↑7/7!+...+(-1)↑(N-1)*X↑(2*N-1)";
150 PRINT "/(2*N-1)!+..."
160 PRINT
170 PRINT "FOR EVERY REAL NUMBER X."
180 PRINT
190 PRINT
200 PRINT
210 PRINT " ","SIN(X) BY"
220 PRINT " ","BUILT IN","SIN(X) BY"
230 PRINT "X","FUNCTION","POWER SERIES"
240 PRINT
250 P=3.14159
260 FOR X=-5*P TO 5*P+0.9999
270     PRINT X,SIN(X),FNS(X)
280 NEXT X
290 DEF FNS(X)      'SIN(X) BY POWER SERIES
300     REM ---REDUCE X TO VALUE NOT GREATER IN MAGNITUDE THAN P/2--
310     X=X-SGN(X)*INT(ABS(X)/(2*P))*2*P
320     IF ABS(X)<=P/2 THEN 350
330     X=SGN(X)*(P-ABS(X))
340     REM---------COMPUTE SIN(X)----------
350     X2=-X↑2
360     I=1
370     FNS=T=X
380     I=I+2
390     T=T*X2/(I*(I-1))     'COMPUTE VALUE OF EACH TERM
400     FNS=FNS+T            'ACCUMULATE SUM OF TERMS IN POWER SERIES
410     IF ABS(T)>=1E-9 THEN 380
420 FNEND
430 END
```

OUTPUT:

THIS PROGRAM ILLUSTRATES THE USE OF A POWER SERIES TO
DEFINE THE SINE FUNCTION. THE FUNCTION SIN[X] IS EQUAL
TO THE POWER SERIES:

    X-X↑3/3+X↑5/5-X↑7/7+...+[-1]↑[N-1]*X↑[2*N-1]/[2*N-1]+...

FOR EVERY REAL NUMBER X.

X	SIN[X] BY BUILT IN FUNCTION	SIN[X] BY POWER SERIES
-15.7079	-1.32723E-05	0
-14.7079	-.841478	-.841471
-13.7079	-.909292	-.909296
-12.7079	-.141107	-.141117
-11.7079	.756811	.756806
-10.7079	.958921	.958923
-9.70795	.279403	.27941
-8.70795	-.656997	-.656991
-7.70795	-.989356	-.989357
-6.70795	-.412106	-.412111
-5.70795	.544032	.54403
-4.70795	.99999	.99999
-3.70795	.536562	.536564
-2.70795	-.420179	-.420177

-1.70795	-.990609	-.990609
-.70795	-.650278	-.650278
.29205	.287916	.287916
1.29205	.961401	.961401
2.29205	.750978	.750977
3.29205	-.14989	-.149893
4.29205	-.912951	-.912952
5.29205	-.836648	-.836647
6.29205	8.86458E-03	8.86989E-03
7.29205	.846227	.84623
8.29205	.905573	.905569
9.29205	.132339	.132331
10.2921	-.762567	-.762572
11.2921	-.956372	-.95637
12.2921	-.270893	-.270885
13.2921	.663644	.663652
14.2921	.98803	.988028
15.2921	.404026	.404013
16.2921	-.551438	-.551449

APPENDIX F

Probability Programs

Program 26: Binomial Experiment

```
100 PRINT TAB(20);"BINOMIAL EXPERIMENT"
110 PRINT "THIS PROGRAM CALCULATES THE PROBABILITY OF SUCCESS OF AN"
120 PRINT "EVENT, GIVEN THE NUMBER OF ATTEMPTS AND THE PROBABILITY"
130 PRINT "OF SUCCESS OF AN ATTEMPT.  FOR EXAMPLE, IF YOU WANT THE"
140 PRINT "PROBABILITY OF GETTING EXACTLY 7 HEADS IN 10 THROWS OF"
150 PRINT "A DIE, ENTER 7, 10, AND 0.5 WHEN THE COMPUTER ASKS."
160 PRINT
170 PRINT
180 PRINT "NUMBER OF SUCCESSES IN EVENT";
190 INPUT M
200 PRINT "NUMBER OF ATTEMPTS";
210 INPUT N
220 PRINT "PROBABILITY OF SUCCESS OF AN ATTEMPT";
230 INPUT P
240 PRINT
250 PRINT "THE PROBABILITY OF THE EVENT IS"FNB(P,N,M)
260 PRINT
270 PRINT
280 PRINT "DO YOU WISH TO RUN AGAIN";
290 INPUT Q$
300 IF Q$>="Y" THEN 160
310 DEF FNB(P,N,M)=P↑M*(1-P)↑(N-M)*FNF(N,M)/FNF(N-M,1)
320 DEF FNF(A,B)
330     FNF=1
340     FOR X=B+1 TO A
350         FNF=X*FNF
360     NEXT X
370 FNEND
380 END
```

OUTPUT:

```
                    BINOMIAL EXPERIMENT
THIS PROGRAM CALCULATES THE PROBABILITY OF SUCCESS OF AN
EVENT, GIVEN THE NUMBER OF ATTEMPTS AND THE PROBABILITY
OF SUCCESS OF AN ATTEMPT.  FOR EXAMPLE, IF YOU WANT THE
PROBABILITY OF GETTING EXACTLY 7 HEADS IN 10 THROWS OF
A DIE, ENTER 7, 10, AND 0.5 WHEN THE COMPUTER ASKS.
```

```
NUMBER OF SUCCESSES IN EVENT ?13
NUMBER OF ATTEMPTS ?30
PROBABILITY OF SUCCESS OF AN ATTEMPT ?0.5

THE PROBABILITY OF THE EVENT IS 0.111535


DO YOU WISH TO RUN AGAIN ?NO
```

APPENDIX G
Special Programs

Program 27: Arranging of Words in Alphabetical Order

```
100 DIM A$(1000)
110 PRINT "THIS PROGRAM ARRANGES WORDS IN ALPHABETICAL ORDER."
120 PRINT
130 PRINT "WHAT WORDS DO YOU WANT ARRANGED";
140 MAT INPUT A$
150 FOR H=1 TO NUM-1
160    FOR I=1 TO NUM-H
170       IF A$(I)<=A$(I+1) THEN 210
180       B$=A$(I)
190       A$(I)=A$(I+1)
200       A$(I+1)=B$
210    NEXT I
220 NEXT H
230 PRINT "YOUR WORDS, IN ALPHABETICAL ORDER, ARE:"
240 MAT PRINT A$,
250 PRINT
260 PRINT
270 PRINT "WOULD YOU LIKE ME TO ARRANGE MORE WORDS";
280 INPUT Q$
290 IF Q$>="Y" THEN 120
300 END
```

OUTPUT:

THIS PROGRAM ARRANGES WORDS IN ALPHABETICAL ORDER.

WHAT WORDS DO YOU WANT ARRANGED ?ONE, TWO, THREE, FOUR
YOUR WORDS, IN ALPHABETICAL ORDER, ARE:

FOUR   ONE   THREE   TWO

WOULD YOU LIKE ME TO ARRANGE MORE WORDS ?YES

```
WHAT WORDS DO YOU WANT ARRANGED ?FIRST, SECOND, THIRD, FOURTH, FIFTH
YOUR WORDS, IN ALPHABETICAL ORDER, ARE:

FIFTH           FIRST           FOURTH          SECOND          THIRD


WOULD YOU LIKE ME TO ARRANGE MORE WORDS ?YES

WHAT WORDS DO YOU WANT ARRANGED ?ARCSIN, ARCCOS, ARCTAN, ARCCOT, ARCCSC
YOUR WORDS, IN ALPHABETICAL ORDER, ARE:

ARCCOS          ARCCOT          ARCCSC          ARCSIN          ARCTAN


WOULD YOU LIKE ME TO ARRANGE MORE WORDS ?NO
```

Program 28: Electronic Configuration of any Element

```
100 DIM E$(24),E(24),A$(18)
110 PRINT "THIS PROGRAM DERIVES THE ELECTRONIC CONFIGURATION OF ANY"
120 PRINT "ELEMENT ACCORDING TO THE ATOMIC ORBITAL THEORY."
130 PRINT
140 REM-----READING ORBITALS AND NUMBER OF ELECTRONS IN ORBITALS-----
150 MAT READ E$,E,A$
160 PRINT
170 PRINT
180 PRINT "WHAT IS THE ATOMIC NUMBER OF YOUR ELEMENT";
190 INPUT N
200 REM----------DERIVING ELECTRONIC CONFIGURATION----------
210 IF N<1 THEN 180
220 IF N<169 THEN 250
230 PRINT "SORRY, I CAN'T HANDLE ELEMENTS THAT BIG."
240 GOTO 180
250 PRINT
260 N=INT(N)
270 IF N<104 THEN 310
280 PRINT "NO SUCH ELEMENT IS KNOWN TO EXIST.  IF IT DID EXIST, ITS"
290 PRINT "ELECTRONIC CONFIGURATION WOULD BE:"
300 GOTO 320
310 PRINT "THE ELECTRONIC CONFIGURATION IS:"
320 S=0
330 FOR I=1 TO 500
340     S=S+E(I)
350     IF S>=N THEN 400
360     PRINT E$(I);A$(E(I));" ";
370     IF I<>16 THEN 390
380     PRINT
390 NEXT I
400 PRINT E$(I);A$(E(I)+N-S)
410 PRINT
420 REM--------------REPEAT ?--------------
430 PRINT
```

```
440 PRINT
450 PRINT "DO YOU WANT ME TO DO ANOTHER ELEMENT";
460 INPUT Q$
470 IF Q$>="Y" THEN 160
480 REM-PRINCIPLE QUANTUM NO., ORBITAL, NO. OF ELECTRONS IN ORBITAL-
490 DATA "1S",2,"2S",2,"2P",6,"3S",2,"3P",6,"4S",2,"3D",10,"4P",6,"5S"
500 DATA 2,"4D",10,"5P",6,"6S",2,"4F",14,"5D",10,"6P",6,"7S",2,"5F"
510 DATA 14,"6D",10,"7P",6,"8S",2,"5G",18,"6F",14,"7D",10,"8P",6
520 DATA "1","2","3","4","5","6","7","8","9","10","11","12","13"
530 DATA "14","15","16","17","18"
540 END
```

OUTPUT:

```
THIS PROGRAM DERIVES THE ELECTRONIC CONFIGURATION OF ANY
ELEMENT ACCORDING TO THE ATOMIC ORBITAL THEORY.


WHAT IS THE ATOMIC NUMBER OF YOUR ELEMENT ?10

THE ELECTRONIC CONFIGURATION IS:
1S2 2S2 2P6


DO YOU WANT ME TO DO ANOTHER ELEMENT ?YES

WHAT IS THE ATOMIC NUMBER OF YOUR ELEMENT ?40

THE ELECTRONIC CONFIGURATION IS:
1S2 2S2 2P6 3S2 3P6 4S2 3D10 4P6 5S2 4D2


DO YOU WANT ME TO DO ANOTHER ELEMENT ?YES

WHAT IS THE ATOMIC NUMBER OF YOUR ELEMENT ?103

THE ELECTRONIC CONFIGURATION IS:
1S2 2S2 2P6 3S2 3P6 4S2 3D10 4P6 5S2 4D10 5P6 6S2 4F14 5D10 6P6 7S2
5F14 6D1


DO YOU WANT ME TO DO ANOTHER ELEMENT ?YES

WHAT IS THE ATOMIC NUMBER OF YOUR ELEMENT ?168

NO SUCH ELEMENT IS KNOWN TO EXIST.  IF IT DID EXIST, ITS
ELECTRONIC CONFIGURATION WOULD BE:
1S2 2S2 2P6 3S2 3P6 4S2 3D10 4P6 5S2 4D10 5P6 6S2 4F14 5D10 6P6 7S2
5F14 6D10 7P6 8S2 5G18 6F14 7D10 8P6


DO YOU WANT ME TO DO ANOTHER ELEMENT ?NO
```

Program 29: Game of Chinese War

```
100 PRINT "THIS PROGRAM PLAYS THE GAME OF CHINESE WAR.  ASSUME"
110 PRINT "THAT THERE ARE 25 STICKS IN A PILE.   THE PLAYERS TAKE"
120 PRINT "TURNS REMOVING 1, 2, 3, OR 4 STICKS FROM THE PILE.  THE"
130 PRINT "PLAYER WHO IS FORCED TO TAKE THE LAST STICK LOSES."
140 MAT READ C$(5),L$(5),W$(5)
150 RANDOM
160 PRINT
170 PRINT
180 PRINT
190 PRINT "DO YOU WANT TO GO FIRST";
200 INPUT Q$
210 PRINT
220 PRINT
230 X=25
240 IF Q$>="Y" THEN 300
250 PRINT "I CHOOSE 4 STICKS."
260 PRINT "THERE ARE NOW 21 STICKS LEFT."
270 X=21
280 PRINT
290 PRINT
300 FOR I=1 TO 10
310     GOSUB 380
320     IF X=1 THEN 650 'OPPONENT WINS
330     IF X<1 THEN 760 'I WIN
340     GOSUB 490
350     IF X=1 THEN 760 'I WIN
360 NEXT I
370 REM----------NUMBER OF STICKS OPPONENT TAKES----------
380 PRINT "HOW MANY STICKS DO YOU WANT";
390 INPUT N
400 IF N<>INT(N) THEN 450
410 IF N<1 THEN 450
420 IF N>4 THEN 450
430 X=X-N
440 RETURN
450 PRINT C$(INT(RND(X)*5+1))
460 PRINT
470 GOTO 380
480 REM----------NUMBER OF STICKS I TAKE----------
490 N=X-5*INT((X-1)/5)-1
500 IF INT((X-1)/5)<>(X-1)/5 THEN 520
510 N=INT(RND(X)*4+1)
520 X=X-N
530 IF N>1 THEN 560
540 PRINT "I'LL TAKE 1 STICK."
550 GOTO 600
560 PRINT "I CHOOSE"N;"STICKS."
570 IF X>1 THEN 600
580 PRINT "THERE'S 1 STICK LEFT."
590 GOTO 610
600 PRINT "THERE ARE NOW"X;"STICKS LEFT."
610 PRINT
620 PRINT
630 RETURN
640 REM----------HE WINS----------
650 PRINT L$(INT(RND(X)*5+1))
660 PRINT
670 PRINT
680 PRINT
690 PRINT
```

```
700 PRINT "LET'S PLAY ANOTHER GAME."
710 PRINT "SINCE YOU WENT FIRST LAST TIME, I'LL GO FIRST."
720 PRINT
730 PRINT
740 GO TO 250
750 REM---------I WIN----------
760 PRINT W$(INT(RND(X)*5+1))
770 F=F+1
780 IF F>3 THEN 950
790 PRINT
800 PRINT
810 PRINT "DO YOU WANT TO PLAY ANOTHER GAME";
820 INPUT R$
830 IF R$>="Y" THEN 160
840 REM---REMARKS IF OPPONENT TAKES ILLEGAL NUMBER OF STICKS---
850 DATA I THINK YOU ARE TRYING TO CHEAT!
860 DATA "I CAN'T ALLOW THAT", DID YOU READ THE DIRECTIONS?
870 DATA "DON'T TRY THAT AGAIN!", "WHY DON'T YOU GIVE UP?"
880 REM---------REMARKS IF OPPONENT WINS----------
890 DATA HOW LUCKY CAN YOU GET!, "YOU'RE BETTER THAN I EXPECTED."
900 DATA "I WASN'T PAYING ATTENTION."
910 DATA "I CAN BEAT YOU, TOO! WATCH THIS.", "NOBODY'S PERFECT"
920 REM---------REMARKS IF I WIN----------
930 DATA THE GOOD GUYS WIN AGAIN!, I GUESS I WON., YOU LOST!
940 DATA I MUST BE SUPERIOR., "NOW DON'T GET ANGRY."
950 END
```

OUTPUT:

```
THIS PROGRAM PLAYS THE GAME OF CHINESE WAR.  ASSUME
THAT THERE ARE 25 STICKS IN A PILE.  THE PLAYERS TAKE
TURNS REMOVING 1, 2, 3, OR 4 STICKS FROM THE PILE.  THE
PLAYER WHO IS FORCED TO TAKE THE LAST STICK LOSES.


DO YOU WANT TO GO FIRST ?YES

HOW MANY STICKS DO YOU WANT ?4
I CHOOSE 4 STICKS.
THERE ARE NOW 17 STICKS LEFT.

HOW MANY STICKS DO YOU WANT ?1
I CHOOSE 4 STICKS.
THERE ARE NOW 12 STICKS LEFT.

HOW MANY STICKS DO YOU WANT ?1
I CHOOSE 2 STICKS.
THERE ARE NOW 9 STICKS LEFT.

HOW MANY STICKS DO YOU WANT ?3
I CHOOSE 2 STICKS.
THERE ARE NOW 4 STICKS LEFT.

HOW MANY STICKS DO YOU WANT ?3
YOU'RE BETTER THAN I EXPECTED.
```

```
LET'S PLAY ANOTHER GAME.
SINCE YOU WENT FIRST LAST TIME, I'LL GO FIRST.


I CHOOSE 4 STICKS.
THERE ARE NOW 21 STICKS LEFT.


HOW MANY STICKS DO YOU WANT ? 3
I CHOOSE 2 STICKS.
THERE ARE NOW 16 STICKS LEFT.


HOW MANY STICKS DO YOU WANT ? 4
I'LL TAKE 1 STICK.
THERE ARE NOW 11 STICKS LEFT.


HOW MANY STICKS DO YOU WANT ? 2
I CHOOSE 3 STICKS.
THERE ARE NOW 6 STICKS LEFT.


HOW MANY STICKS DO YOU WANT ? 5
DID YOU READ THE DIRECTIONS ?

HOW MANY STICKS DO YOU WANT ? 0
WHY DON'T YOU GIVE UP ?

HOW MANY STICKS DO YOU WANT ? 1
I CHOOSE 4 STICKS.
THERE'S 1 STICK LEFT.


YOU LOST!


DO YOU WANT TO PLAY ANOTHER GAME ? NO
```

Program 30: A Meaningless Technical Report

```
1000 REM SOURCE FOR MOST OF THE PHRASES: RAYMOND A. DEFFRY, "NEW
1010 REM S-I-M-P PROGRAM MAKES TECH WRITING EASY AS 1-2-3-4-5-6..."
1020 REM IN SOFTWARE AGE, VOLUME 2-NO. 11, DECEMBER, 1968, PAGES 25-26
1030 REM         THIS PROGRAM WRITES MEANINGLESS TECHNICAL PAPERS BY
1040 REM COMBINING RANDOM PHRASES USED IN THE AEROSPACE INDUSTRY.
1050 REM ONE PHRASE IS TAKEN FROM EACH OF THE FOUR GROUPS TO FORM
1060 REM IMPRESSIVE, GRAMMATICALLY CORRECT SENTENCES.
1070 DIM A$(20),A(300),B(70),C(70),D(70)
1080 PRINT "HOW MANY PAGES OF INFORMATION DO YOU WANT";
1090 INPUT P
1100 PRINT
1110 PRINT
1120 PRINT "    THE ECONOMIC CONSIDERATIONS OF THE AEROSPACE INDUSTRY"
1130 PRINT
1140 PRINT
```

Appendix G: Special Programs   337

```
1150 RANDOM
1160 MAT READ A$, B$, C$, D$
1170 PRINT "          ";
1180 K=1        'NUMBER OF SPACES USED IN LINE
1190 F=13       'NUMBER OF LINES PRINTED
1200 REM----------CHOOSING RANDOM PHRASES------------
1210 X1=INT(RND(X)*20+1)
1220 IF F>13 THEN 1240
1230 X1=0
1240 X2=INT(RND(X)*10+1)
1250 X3=INT(RND(X)*10+1)
1260 X4=INT(RND(X)*10+1)
1270 IF RND(X)>=1/2 THEN 1340
1280 F$=B$(X2)
1290 G$=D$(X4)
1300 X2=X4=0
1310 B$(0)=G$
1320 D$(0)=F$
1330 REM--COMBINING PHRASES USING THE CODE NUMBERS OF EACH LETTER--
1340 CHANGE A$(X1) TO A
1350 CHANGE B$(X2) TO B
1360 A(A(0)+1)=44       'COMMA
1370 IF F>13 THEN 1390
1380 A(A(0)+1)=0
1390 L=L1=A(0)+2
1400 GOSUB 1520
1410 L=L2=L1+B(0)+1
1420 CHANGE C$(X3) TO B
1430 GOSUB 1520
1440 L=L3=L2+B(0)+1
1450 CHANGE D$(X4) TO B
1460 GOSUB 1520
1470 L4=L3+B(0)+1
1480 A(L4)=46           'PERIOD
1490 A(L1)=A(L2)=A(L3)=A(L4+1)=A(L4+2)=32       'SPACES
1500 GOTO 1570
1510 REM-----SUBROUTINE WHICH FILLS ARRAY A WITH CHARACTERS OF PHRASES
1520 FOR I=1 TO B(0)
1530    A(I+L)=B(I)
1540 NEXT I
1550 RETURN
1560 REM--PRINTING SENTENCE AND LINE FEEDING AT END OF WORD--
1570 P(0)=I=1
1580 FOR K=K TO 72
1590    IF K<55 THEN 1680
1600    IF K=72 THEN 1620
1610    IF A(I)<>32 THEN 1680
1620    PRINT
1630    F=F+1   'NUMBER OF LINES
1640    K=1     'POSITION ON LINE
1650    IF INT(F/50)=F/50 THEN 1740 'PAGING
1660    IF A(I-1)=46 THEN 1780
1670    GO TO 1720
1680    P(1)=A(I)
1690    CHANGE P TO S$
1700    PRINT S$;
1710    IF I=L4+2 THEN 1780
1720    I=I+1
1730 NEXT K
1740 FOR H=1 TO 20
1750    PRINT
1760 NEXT H
1770 IF A(I-1)<>46 THEN 1720
1780 Z=Z+1     'NUMBER OF SENTENCES
```

```
1790 IF INT(Z/4)<Z/4 THEN 1840 'PARAGRAPHING
1800 IF K=1 THEN 1820
1810 PRINT
1820 PRINT "     ";
1830 K=6
1840 IF ABS(F-50*P)>4 THEN 1210 'NUMBER OF PAGES
1850 PRINT
1860 REM------------PARENTHETICAL PHRASES------------
1870 DATA IN PARTICULAR, ON THE OTHER HAND, HOWEVER, SIMILARLY
1880 DATA AS A RESULTANT IMPLICATION, IN THIS REGARD
1890 DATA BASED ON INTEGRAL SUBSYSTEM CONSIDERATIONS
1900 DATA FOR EXAMPLE, THUS, IN RESPECT TO SPECIFIC GOALS
1910 DATA UTILIZING THE ESTABLISHED HYPOTHESES, MOREOVER, IN ADDITION
1920 DATA IN VIEW OF SYSTEM OPERATION, FURTHERMORE, TO SOME EXTENT
1930 DATA CONSIDERING THE POSTULATED INTERRELATIONSHIPS
1940 DATA FOR THE MOST PART, INDEED
1950 DATA BASED ON SYSTEM ENGINEERING CONCEPTS
1960 REM------------NOUNS AND MODIFIERS------------
1970 DATA A LARGE PORTION OF THE INTERFACE COORDINATION COMMUNICATION
1980 DATA A CONSTANT FLOW OF EFFECTIVE INFORMATION
1990 DATA THE CHARACTERIZATION OF SPECIFIC CRITERIA
2000 DATA INITIATION OF CRITICAL SUBSYSTEM DEVELOPMENT
2010 DATA THE FULLY INTEGRATED TEST PROGRAM
2020 DATA THE PRODUCT CONFIGURATION BASELINE
2030 DATA ANY ASSOCIATED SUPPORTING ELEMENT
2040 DATA THE INCORPORATION OF ADDITIONAL MISSION CONSTRAINTS
2050 DATA THE INDEPENDENT FUNCTION PRINCIPLE
2060 DATA A CONSIDERATION OF SYSTEM AND/OR SUBSYSTEM TECHNOLOGIES
2070 REM------------VERBS AND PREPOSITIONS------------
2080 DATA MUST UTILIZE AND BE FUNCTIONALLY INTERWOVEN WITH
2090 DATA MAXIMIZES THE PROBABILITY OF SUCCESS AND MINIMIZES TIME FOR
2100 DATA ADDS EXPLICIT PERFORMANCE LIMITS TO
2110 DATA NECESSITATES THAT URGENT CONSIDERATION BE APPLIED TO
2120 DATA REQUIRES CONSIDERABLE SYSTEMS ANALYSIS TO ARRIVE AT
2130 DATA "IS FURTHER COMPOUNDED, WHEN TAKING INTO ACCOUNT"
2140 DATA PRESENTS EXTREMELY INTERESTING CHALLENGES TO
2150 DATARECOGNIZES THE IMPORTANCE OF OTHER SYSTEMS AND NECESSITY FOR
2160 DATA EFFECTS A SIGNIFICANT IMPLEMENTATION TO
2170 DATA ADDS OVERRIDING PERFORMANCE CONSTRAINTS TO
2180 REM------------NOUNS AND MODIFIERS------------
2190 DATA THE SOPHISTICATED HARDWARE
2200 DATA THE ANTICIPATED THIRD GENERATION EQUIPMENT
2210 DATA THE SUBSYSTEM COMPATIBILITY TESTING
2220 DATA THE STRUCTURAL DESIGN
2230 DATA THE PRELIMINARY QUALIFICATION LIMIT
2240 DATA THE PHILOSOPHY OF COMMONALITY AND STANDARDIZATION
2250 DATA THE EVOLUTION OF SPECIFICATIONS OVER A GIVEN TIME
2260 DATA THE GREATER FLIGHT-WORTHINESS CONCEPT
2270 DATA ANY DISCRETE CONFIGURATION MADE
2280 DATA THE TOTAL SYSTEM RATIONALE
2290 END
```

OUTPUT:

HOW MANY PAGES OF INFORMATION DO YOU WANT ? 2

THE ECONOMIC CONSIDERATIONS OF THE AEROSPACE INDUSTRY

    ANY ASSOCIATED SUPPORTING ELEMENT REQUIRES CONSIDERABLE SYSTEMS ANALYSIS TO ARRIVE AT THE SUBSYSTEM COMPATIBILITY TESTING. FURTHERMORE, THE PHILOSOPHY OF COMMONALITY AND STANDARDIZATION NECESSITATES THAT URGENT CONSIDERATION BE APPLIED TO THE PRODUCT CONFIGURATION BASELINE. TO SOME EXTENT, THE GREATER FLIGHT-WORTHINESS CONCEPT ADDS EXPLICIT PERFORMANCE LIMITS TO THE CHARACTERIZATION OF SPECIFIC CRITERIA. TO SOME EXTENT, THE STRUCTURAL DESIGN IS FURTHER COMPOUNDED, WHEN TAKING INTO ACCOUNT THE FULLY INTEGRATED TEST PROGRAM.
    AS A RESULTANT IMPLICATION, THE PRODUCT CONFIGURATION BASELINE MUST UTILIZE AND BE FUNCTIONALLY INTERWOVEN WITH THE PHILOSOPHY OF COMMONALITY AND STANDARDIZATION. INDEED, A CONSTANT FLOW OF EFFECTIVE INFORMATION RECOGNIZES THE IMPORTANCE OF OTHER SYSTEMS AND NECESSITY FOR THE STRUCTURAL DESIGN. IN VIEW OF SYSTEM OPERATION, THE ANTICIPATED THIRD GENERATION EQUIPMENT ADDS EXPLICIT PERFORMANCE LIMITS TO THE PRODUCT CONFIGURATION BASELINE. FOR THE MOST PART, THE TOTAL SYSTEM RATIONALE EFFECTS A SIGNIFICANT IMPLEMENTATION TO THE CHARACTERIZATION OF SPECIFIC CRITERIA.
    FOR EXAMPLE, THE STRUCTURAL DESIGN EFFECTS A SIGNIFICANT IMPLEMENTATION TO THE CHARACTERIZATION OF SPECIFIC CRITERIA. IN THIS REGARD, THE PRELIMINARY QUALIFICATION LIMIT RECOGNIZES THE IMPORTANCE OF OTHER SYSTEMS AND NECESSITY FOR A CONSIDERATION OF SYSTEM AND/OR SUBSYSTEM TECHNOLOGIES. BASED ON INTEGRAL SUBSYSTEM CONSIDERATIONS, THE SOPHISTICATED HARDWARE RECOGNIZES THE IMPORTANCE OF OTHER SYSTEMS AND NECESSITY FOR A CONSIDERATION OF SYSTEM AND/OR SUBSYSTEM TECHNOLOGIES. FOR EXAMPLE, THE TOTAL SYSTEM RATIONALE REQUIRES CONSIDERABLE SYSTEMS ANALYSIS TO ARRIVE AT THE INDEPENDENT FUNCTION PRINCIPLE.
    INDEED, THE TOTAL SYSTEM RATIONALE NECESSITATES THAT URGENT CONSIDERATION BE APPLIED TO THE FULLY INTEGRATED TEST PROGRAM. THUS, THE FULLY INTEGRATED TEST PROGRAM EFFECTS A SIGNIFICANT IMPLEMENTATION TO ANY DISCRETE CONFIGURATION MADE. THUS, THE SUBSYSTEM COMPATIBILITY TESTING ADDS EXPLICIT PERFORMANCE LIMITS TO THE CHARACTERIZATION OF SPECIFIC CRITERIA. ON THE OTHER HAND, THE PHILOSOPHY OF COMMONALITY AND STANDARDIZATION PRESENTS EXTREMELY INTERESTING CHALLENGES TO A LARGE PORTION OF THE INTERFACE COORDINATION COMMUNICATION.

TO SOME EXTENT, A LARGE PORTION OF THE INTERFACE COORDINATION COMMUNICATION NECESSITATES THAT URGENT CONSIDERATION BE APPLIED TO THE SOPHISTICATED HARDWARE. INDEED, ANY DISCRETE CONFIGURATION MADE EFFECTS A SIGNIFICANT IMPLEMENTATION TO THE CHARACTERIZATION OF SPECIFIC CRITERIA. IN THIS REGARD, THE PHILOSOPHY OF COMMONALITY AND STANDARDIZATION RECOGNIZES THE IMPORTANCE OF OTHER SYSTEMS AND NECESSITY FOR ANY ASSOCIATED SUPPORTING ELEMENT. IN RESPECT TO SPECIFIC GOALS, THE EVOLUTION OF SPECIFICATIONS OVER A GIVEN TIME RECOGNIZES THE IMPORTANCE OF OTHER SYSTEMS AND NECESSITY FOR A LARGE PORTION OF THE INTERFACE COORDINATION COMMUNICATION.

UTILIZING THE ESTABLISHED HYPOTHESES, THE SUBSYSTEM COMPATIBILITY TESTING MAXIMIZES THE PROBABILITY OF SUCCESS AND MINIMIZES TIME FOR THE INDEPENDENT FUNCTION PRINCIPLE. UTILIZING THE ESTABLISHED HYPOTHESES, INITIATION OF CRITICAL SUBSYSTEM DEVELOPMENT RECOGNIZES THE IMPORTANCE OF OTHER SYSTEMS AND NECESSITY FOR THE ANTICIPATED THIRD GENERATION EQUIPMENT. MOREOVER, THE SOPHISTICATED HARDWARE PRESENTS EXTREMELY INTERESTING CHALLENGES TO A LARGE PORTION OF THE INTERFACE COORDINATION COMMUNICATION. AS A RESULTANT IMPLICATION, INITIATION OF CRITICAL SUBSYSTEM DEVELOPMENT PRESENTS EXTREMELY INTERESTING CHALLENGES TO THE PRELIMINARY QUALIFICATION LIMIT.

FOR EXAMPLE, THE SUBSYSTEM COMPATIBILITY TESTING PRESENTS EXTREMELY INTERESTING CHALLENGES TO A CONSIDERATION OF SYSTEM AND/OR SUBSYSTEM TECHNOLOGIES. THUS, THE TOTAL SYSTEM RATIONALE PRESENTS EXTREMELY INTERESTING CHALLENGES TO THE PRODUCT CONFIGURATION BASELINE. CONSIDERING THE POSTULATED INTERRELATIONSHIPS, THE GREATER FLIGHT-WORTHINESS CONCEPT REQUIRES CONSIDERABLE SYSTEMS ANALYSIS TO ARRIVE AT ANY ASSOCIATED SUPPORTING ELEMENT. IN ADDITION, THE SOPHISTICATED HARDWARE IS FURTHER COMPOUNDED, WHEN TAKING INTO ACCOUNT THE PRODUCT CONFIGURATION BASELINE.

IN THIS REGARD, A CONSIDERATION OF SYSTEM AND/OR SUBSYSTEM TECHNOLOGIES RECOGNIZES THE IMPORTANCE OF OTHER SYSTEMS AND NECESSITY FOR THE EVOLUTION OF SPECIFICATIONS OVER A GIVEN TIME. ON THE OTHER HAND, THE INCORPORATION OF ADDITIONAL MISSION CONSTRAINTS ADDS OVERRIDING PERFORMANCE CONSTRAINTS TO THE GREATER FLIGHT-WORTHINESS CONCEPT. THUS, THE PHILOSOPHY OF COMMONALITY AND STANDARDIZATION EFFECTS A SIGNIFICANT IMPLEMENTATION TO THE INCORPORATION OF ADDITIONAL MISSION CONSTRAINTS. MOREOVER, THE ANTICIPATED THIRD GENERATION EQUIPMENT ADDS OVERRIDING PERFORMANCE CONSTRAINTS TO ANY ASSOCIATED SUPPORTING ELEMENT.

HOWEVER, THE SOPHISTICATED HARDWARE PRESENTS EXTREMELY INTERESTING CHALLENGES TO INITIATION OF CRITICAL SUBSYSTEM DEVELOPMENT. IN RESPECT TO SPECIFIC GOALS, THE PRELIMINARY QUALIFICATION LIMIT MAXIMIZES THE PROBABILITY OF SUCCESS AND MINIMIZES TIME FOR A LARGE PORTION OF THE INTERFACE COORDINATION COMMUNICATION.

INDEX

ABS(X), 45
Addition, 22
Additive inverse of a matrix, 190
Algebra Quiz, 124
Alphabetic Sort, 331
Alphanumeric Data, 71
ALTMODE (ESCAPE or PREFIX) Key, 27
American Standard Code for Information Interchange, 118
Ampersand (&), 114
Analysis of General Quadratic, 308
Angle-Side-Angle, 154
Apostrophe, 111
Arccos, 158
Arcsin, 156
Area of Circle, 62,69,78
Area of Regular Polygon, 301
Area of a Triangle, 304
Area Under Curve of F, 235,244,249,319
Argument of a function, 45
Arithmetic mean, 226
Arithmetic operations, 22
ASCII Code, 118,120
ASR-33 Teletype, 5
Assignment statements, 259
Assignment of values to variables, 29
ATN(X), 46,56
Average of N Numbers, 279
Average Value of a Function, 319

Back Arrow (SHIFT O), 25
BASIC, 2,10
Binomial Experiment, 329
Buffer, input, 8
Bugs, 148
Built-in functions, 45,260
Buzzer Release Button, 5
BYE, 9

CATALOG, 266
Central Processor, 12
CHANGE Command, 118
Characteristics of User-Defined functions, 133

Coefficient of correlation, 228
Columns in BASIC, 14
Commas with numbers (PRINT), 14
Commas with numbers and strings (PRINT), 15
Commas with strings (PRINT), 14
Comments, 111,154,260
Compilation Errors, 270
Compilation of a program, 68,148
Completion Time of Work, 285
CON, 198,201
Configuration of an Element, 332
Constants, 269
Continuous function, 180
CONTROL-SHIFT P, 266
Control Statements, 259
Conversion from Base 10, 286
Corrections, 25,148
Correlation coefficient, 228
COS(X), 46,55
Cosecant function, 56
COT(X), 46,56
Counter, 86

DATA pointer, 38
DATA restrictions, 269
DATA Statement, 35
    Excess of DATA, 38
    Insufficient DATA, 38
    Legal Numbers in, 37
    Placement of DATA, 35
    Strings, 73
Date, 11
Debugging, 117,148
Decimals, 58
Decision box, 61,69
DEF Command, 131
    Characteristics, 133
    Multiline functions, 134
DEF FN*(X), 131
Defining functions, 130
Degree measure, 170
DELETE, 267
DELETED, 27

Determinant of a matrix, 194
Diagram, 169
Dice Game, 107
Difference of matrices, 190,204
DIM Command, 101,197
    Matrices, 197
    Numerical variables, 102
    Placement of DIM, 103
    Several variables, 103
    String variables, 104
Dimension of a matrix, 189, 210
Disc, 65
Distinction between GOSUB and GO TO
    Commands, 110
Division, 22
Division of Polynomials, 290
Dollar sign following variable name, 71
Domain, 45

Echo, 6
EDIT Commands, 267
END Statement, 12
End of Transmission, 122
Entry of a matrix, 188
Equal to, 67
Equality of matrices, 189,208
Equation of Parabola, 306
Equation of regression line, 229
Error Messages, 21,25,148,270
Escape Key, 27
Executive Errors, 273
Execution of a program, 11,64
Exponential notation, 19
Exponentiation, 22
Expression, 258
EXP(X), 46,54
EXTRACT, 267
Extracting characters from a string, 122

Factorial (N!), 87,99,136
FDX button, 6
Flag, 84,235
Flow chart, 60
FN*(X), 131
FNEND, 134
FOR-NEXT Command, 91
Full Duplex button, 5
Function, 45,130
    Absolute value, 47
    Exponential, 54
    Greatest Integer, 47
    Logarithmic, 52
    Random Number, 49
    Sign, 47
    Square Root, 51
    Trigonometric, 54
    User Defined, 131

Function End Statement, 134
Functions and subroutines, 138

Game of Chinese War, 334
General techniques of debugging, 154
GOSUB-RETURN Commands, 107
GO TO Command, 79
Graph by computer, 173
Graph of any function, 171
Graph of F with X-Axis Horizontal, 292
Graph Up to 26 Functions, 293
Greater than, 67
Greatest Common Factor, 282
Greatest Integer Function, 46

Hardware, 1
Heading, 11
HERE IS Key, 4

Identity matrix, 193,198,201
IDN, 198,201
IF-THEN Command, 67
    with string variables, 75
ILLEGAL INSTRUCTION IN N, 149
Illegal MAT statements, 210
Image set, 45
Increment value, 176
Initializing counters, 86
Input buffer storage area, 8
INPUT Command, 41
    Legal Numbers, 43
    Several Variables, 41
    Strings, 73
Inserting steps, 12
Insufficient DATA, 38
Integers, 58
INT(X), 46
INV(A), 198
INVALID USER, 6
Inverse of a matrix, 193

KEY, 8

Law of Cosines, 154
Law of Sines, 154
LCL button, 4
Least Common Multiple, 282
Legal Expressions in LET Statements, 31
Legal Expressions in PRINT Statements, 19
Legal names for variables, 31
Legal nested loops, 97
Legal numbers in DATA Statements, 37
Legal numbers in INPUT Statements, 43
LENGTH, 265
Less than, 67
LET Command, 29
    Legal Expressions, 31

Index   343

Omitting word LET, 34
Same value to several variables, 33
Same variable on both sides of equal
   sign, 33
Strings, 71
Limit of a Function, 314
LINE FEED Key, 4
Line numbers, 10
LIST, 65,265
   LIST--N, 265
   LISTNH, 265
   LISTNH--N, 265
Lists and Tables (Vectors and Matrices),
   restrictions, 269
LOCAL Button, 5
LOCAL Mode, 3
LOGging OFF, 8
Logical progression ERROR, 152
LOG(N) in Base B, 288
LOG ON, 4,6
LOGOUT, 9
LOG(X), 46,52
Loop, 85,91
   nested, 97
   transfer of control outside, 98

Machine Language Instructions, 148
MAT INPUT Command, 113,198,222
MAT PRINT Statement, 197
MAT READ Statement, 197
MAT Statements, 196
   CON, 201
   DET, 205
   Difference, 204
   Equality, 207
   IDN, 201
   Illegal, 210
   INPUT, 113,222
   INV(A), 205
   PRINT, 197
   Product, 205
   READ, 197
   Same variable on both sides of equal
      sign, 208
   Scalar Product, 204
   Strings, 216
   Sum, 204
   TRN(A), 205
   ZER, 201
Matrices, 188
   Additive inverse, 190
   Difference, 190
   Identity, 193
   Inverse, 193
   Product, 191
   Scalar product, 189

Solution of linear equations, 195
Sum, 189
Transpose, 191
Zero, 190
Matrix equality, 207
Matrix of ones, 198,201
Matrix solution of linear systems, 220
Mean, Arithmetic, 226
Meaningless Technical Report, 336
MERGE, 267
Midpoint Rule, 240
Mistakes, 148
Modulo 75,18
MOVE, 268
Multiline function definitions, 134
Multiple-way conditional switch, 83
Multiplication, 22
Multiplication symbol, 14

Name, program, 11
N-dimensional arrays, 212
Negative sign, 17
Nested loops, 97
Nested subroutines, 112,128
NEW, 264
NEW OR OLD--, 7
NEW PROGRAM NAME--, 7
Normal distribution curve, 179,249
Not equal to, 67
NUM, 113,223
Number, 258
Numbers in BASIC, 57
Numerical variable, 19,258

OFF LINE Mode, 3
OLD, 66,264
OLD PROGRAM NAME--, 66
ON-GO TO Command, 83
ON LINE Mode, 3,7
ON-THEN Command, 85
Operations, 22,262
Order of a matrix, 193
Order of operations, 22
Ordered pairs (x,y), 171
ORIG Button, 5,6
OUTPUT, 12,260
OUTPUT numbers, 58

PAGE, 268
Pairs of Factors, 280
Parentheses, 23
Percentage of variance, 229
PREFIX Key, 27
Pre-image set, 45
Prime Factors, 281
PRINT Command, 10
   Alone, 16

PRINT Command (cont.)
    Commas with Numbers, 14
    Commas with Numbers and Strings, 15
    Commas with Strings, 14
    Legal Expressions, 19
    Numbers, 11
    Semicolon with Numbers, 16
    Semicolon with Numbers and Strings, 42
    Semicolon with Strings, 17
    Strings, 13
    TAB(N), 18
Product of matrices, 191,205
Program, 10
Program length restriction, 269
Program name, 11
Programmer-defined functions, 131
Punching a tape, 3,64
Pythagorean Triplets, 300

Quotation marks, 13

Radian, 55
Radian measure, 170
RANDOM, 50
RANDOMIZE, 50
Random Number Function, 46
Range, 45
READ Command, 35
    Order of assignment of values, 35
    Placement of READ, 37
    RESTORE, 38,75
    Strings, 73
Reading in a punched tape, 7
READY, 7
Real Roots of $AX^2+BX+C = 0$, 79
Real Zeros of a Function, 180
Real Zeros by Newton's Method, 297
Rectangular coordinate system, 171
Rectangular Rule, 236
Regression line, 229
Regular Polygon, Interior Angle and Diagonals, 302
Relations in BASIC, 67
Relative error, 185
REM Statement, 111
RENAME, 266
Repetition, 78
REPLACE, 265
Replacing a line, 12,26
RESEQUENCE, 268
RESTORE Command, 38
RESTORE* Command, 75
RESTORE$ Command, 75
RETURN Command, 107
RETURN Key, 4,11
Retyping a line, 12,26

RND(X), 46
Root search, 183
Rounding off, 47
Round-off Error, 146,170
Routines, 117
Rth Root of N, 290
RUB-OUT, 122
RUN, 11,265
RUNOFF, 268

Same matrix on both sides of equal sign, 208
Same variable on both sides of equal sign, 33
SAVE, 265
Scalar product of matrices, 189,204
SCRATCH, 266
Secant function, 56
Secant line, 141
Semicolon with numbers (PRINT), 17
Semicolon with numbers and strings (PRINT), 42
Semicolon with strings (PRINT), 17
SGN(X), 46
SHIFT O Key (Back Arrow), 25
Side-Angle-Side, 155
Side-Side-Angle, 159
Side-Side-Side, 157
Sigma, 227
Simpson's Rule, 242
SIN(X), 46,55
SIN(X) by Power Series, 326
Size restrictions, 269
Skipping a column, 14
Skipping a line, 16
Slope of F(X) at X = C, 139
Software, 1
Solution of $AX+B = CX+D$, 279
Solution of linear equations, 195
Solution of Linear Systems, 218,284
Solution of Triangles, 154
SPACE BAR, 4
Spaces, 11
SQR(X), 46,289
Standard deviation, 227
Statements, 259
Statistical Package, 225,231
Step numbers, 10
STOP, 266
Store, 11
String, 13,71
String manipulation, 71
String Statements, 71
String variables, 19,71
    comparison of, 76
    mixing numeric and string data, 74
    renaming a string, 72
Strings with MAT Statements, 216

Subroutine, 107
Subscripted variables, 32,101,258
Subtraction, 22
Sum, Product, Maximum, Minimum, 114
Sum of matrices, 189,204
Symbols in arithmetic operations, 22
Syntax, 148
SYSTEM, 7
System Commands, 264

TAB(N), PRINT, 18
Tangent line, 141
TAN(X), 46,55
TAPE, 8,266
Tape Punch controls, 5
Tape Reader, 8
Tape Reader controls, 5
Taylor series for Sin(X), 256,326
Techniques of Debugging, 148
Teletype (ASR-33,35,37), 1,3
Time, 11
Time-sharing system, 1
Transpose of a matrix, 191,198,205
Trapezoidal Rule, 238
Triangle solution program, 164
Trigonometric functions, 54

TRN(A), 198,205
Turn-around time, 1

Undefined variables, 19,21,32,86
UNSAVE, 266
User-defined functions, 131

Value of a function, 45,130
Variables, 29
    assignment of same value to several, 33
    legal names, 31
    same on both sides of equal sign, 33
Variance, 227
Vectors, 210
Volume of a Solid of Revolution, 322
Volume of Sphere, 62,69,78

WEAVE, 269
Working storage area, 8

ZER, 198,201
Zeros of a Function, 180
Zero matrix, 190,198,201